SEARCHING FOR HOME WATERS

Searching for Home Waters

A BROOK TROUT PILGRIMAGE

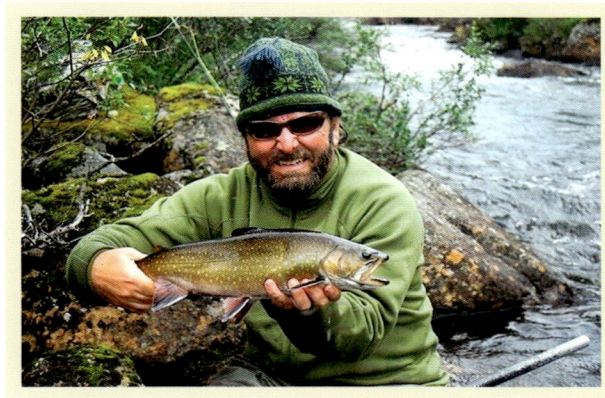

Michael K. Steinberg

with original watercolors by Karen Talbot and original drawings by Frederick Steinberg

PUBLISHED IN ASSOCIATION WITH

George F. Thompson Publishing and the Center for the Study of Place

The University of Georgia Press   *Athens*

SOUTHERN HIGHLANDS RESERVE
Publications

Searching for Home Waters: A Brook Trout Pilgrimage was developed by George F. Thompson Publishing (www.gftbooks.com) for publication by the University of Georgia Press.

Project Director and Editor: George F. Thompson
Editorial and Research Assistant: Mikki Soroczak
Manuscript Editor: Purna Makaram

The publisher extends special gratitude to Karen Talbot and Frederick Steinberg, for providing original art to this book.

Designed by Erin Kirk
Set in Miller Text
Printed and bound by Friesens

The paper in this book meets the guidelines for permanence and durability of the Committee on Production Guidelines for Book Longevity of the Council on Library Resources.

Most University of Georgia Press titles are available from popular e-book vendors.

Printed in Canada
23 24 25 26 27 C 5 4 3 2 1

Library of Congress Cataloging-in-Publication Data
Names: Steinberg, Michael K., 1965– author. | Talbot, Karen (illustrator), illustrator. | Steinberg, Frederick, illustrator.
Title: Searching for home waters : a brook trout pilgrimage / Michael K. Steinberg ; with original watercolors by Karen Talbot and original drawings by Frederick Steinberg.
Description: Athens : The University of Georgia Press, [2023] | "Published in association with George F. Thompson Publishing and the Center for the Study of Place"— Title page. | Includes bibliographical references.
Identifiers: LCCN 2022053987 | ISBN 9780820363622 (hardback) | ISBN 9780820363639 (epub) | ISBN 9780820364216 (pdf)
Subjects: LCSH: Brook trout fishing—East (U.S.) | Brook trout—Geographical distribution—Climatic factors—East (U.S.) | Brook trout—Ecology—East (U.S.)
Classification: LCC SH689.3 .S74 2023 | DDC 597.5/540275—dc23/eng/20221212
LC record available at https://lccn.loc.gov/2022053987

TITLE PAGE: A selfie of me holding a nice brookie in Labrador, Canada, 2010.

FOR FREDERICK AND TRUMAN

I once fished a river in Connecticut fed by many springs. It headed into a tangled field of deadfall and thick grasses, a swampy place through which ran rivulets of cold, crystalline water dimpled everywhere with the rises of little fish. The rises were all from wild brookies. They were no more than four or five inches long—sleek, velvety fish with black, mottled backs and jewel-like, bright-red spots. When I hooked them on a Hairwing Royal Coachman, they seemed to pop directly up out of the water, wriggling vigorously. I marveled at their beauty, and each of them made me smile. We were not ten miles from a suburban world peppered with strip malls, used-car lots, gaudy gas stations, and fast-food restaurants, but this was a place exquisitely separate, quiet, and wild, and I was hugely grateful for it, for the rich satisfaction it brought me, for the wild brookies that are the emblem of such increasingly rare wildness.

NICK LYONS TO THE AUTHOR, 2009

Contents

Bug List

The Wild River in western Maine, one of my favorite home waters. Photograph by Michael K. Steinberg, 2015.

Preface

AFTER FINISHING MY FIRST BOOK, *Stalking the Ghost Bird* (2008), I began to outline my next major project and was soon offered a contract by a top university press to write a book about the restoration of rivers in New England. That project made sense because of the growing interest in removing dams in the region, and I spend my summers in Maine, close to the Penobscot and Kennebec Rivers whose restoration has garnered a great deal of attention. The larger river restoration issue was front and center in not only my mind but in the larger environmental/conservation community as well.

With a contract in hand, I began the process of identifying which rivers I could write about. That was as far as I got, for I realized that, while I was emotionally connected to rivers in a general sense as an avid fly angler, no major theme or issue emerged to inspire enough passion in me to complete a book. Even as I recognized the ecological importance of restoring various large New England rivers, they weren't *my* rivers. I neither grew up on their shores nor paddled their waters, so there was no tangible

personal history. I tore up my contract and started over. This book is the result of that process.

It began when I contemplated what it is about rivers that initially attracted me to the topic. Why was I feeling compelled to consider writing such a book? The answer now is obvious: I wanted to write about rivers so I could spend time on them, fly-fishing for trout, specifically for brook trout. I am, of course, interested in the plight of rivers—how could I not be as an angler who knows that brook trout tell us a great deal about the health of the larger environment, as they are an indicator species for streams, lakes, and watersheds? But brook trout are really the topic that drives my interest in rivers and streams.

I am not alone in my interest in the conservation of brook trout and their environments. Today brook trout receive more conservation attention than they ever have, thanks to growing numbers of conservation and restoration projects. The impetus for this attention is the Eastern Brook Trout Joint Venture, an association of state fish and wildlife agencies, local and federal resource agencies, academic institutions, and private-sector conservation organizations, such as Trout Unlimited, that are working to conserve eastern brook trout mainly through the restoration of habitat and other improvement projects. During my travels, I visited many individuals and projects associated with the joint venture, and I describe some of them in this book.

Initially, I resolved to visit the entire range of the "brookie" in eastern North America, including every state that has projects designed to restore or protect the populations of brook trout, as well as Labrador, the Canadian province that is home to the biggest brook trout in the world. Thus, beginning in northern Georgia and northwestern South Carolina, I basically followed the range of the brook trout along the spine of the Appalachian Mountains before ending my journey in Labrador. The overall journey took four years. Since I never intended to visit every restoration project in every state, I decided to highlight specific projects, people, and environmental issues I encountered along the way. But beyond highlighting conservation efforts at the state level, I also explore why the brook trout has become so important to me and why this trip felt like a pilgrimage, a journey of the heart, rather than a fishing excursion. So this is not a "how-to" book. I do not often offer tips on angling in the various North American rivers and streams, but I do share my connection with this special fish and its landscapes, a connection so deep that has it has altered my professional and personal life in significant ways.

For many—dare I say most?—dedicated fly anglers, the personal and natural worlds comingle closely. Our personal lives are wrapped up in and around "our" rivers, streams, brooks, and flats. I cannot separate the personal from the natural. The idea of "home waters" is a popular theme in fly-fishing. The very idea of home and water shows succinctly why the personal and natural are nearly inseparable for fly-fishers. The environments where I fish reflect critical parts of my personality, my family life, and my professional life. I have met countless other fly anglers during my travels who share this close connection to a home water and the natural world. So I cannot tell the story of a stream or fish without also describing my own story and journey.

Why one person chooses to fish for trout and another targets bass is the result of myriad personal decisions, attitudes, and geography. We see and experience nature and angling through our own actions, values, and backgrounds. For me, brook trout mean something special. And that something is clean cold water in remote areas, away from crowds and hatchery-stocking trucks. I can't describe that environment and experience without first describing what the species and environment mean to me.

As a geographer, I like to organize information spatially. This book is no exception. Because my travels initially took me from south to north, the chapters follow a similar pattern. This does not match the chronological order of my travels, but it made intuitive sense to follow the natural geography of the Appalachian Mountains—from the Deep South to New England—and then into eastern Canada. Another note on geography: I have altered or omitted some names of specific streams I visited. This is not to mislead the reader but to keep the names of specific streams ambiguous, at the request of state officials. When I discuss well-known streams or well-publicized restoration projects, I make no attempt to hide specifics. These streams are already fished hard or covered so extensively that my naming them won't lead to new crowds to their waters.

Acknowledgments

I WANT TO THANK the following individuals whose advice, contributions, and support made it possible for me to write this book: Nick Lyons, whose words in email, by telephone, and in his own writings were inspiring; Dave Van Lear in South Carolina; Leon Brotherton in Georgia; Jim Herrig, Matt Kulp, and Ian Rutter in Tennessee; Doug Bessler in North Carolina; David Demarest, Urbie Nash, John Ross, Larry Mohn, and Seth Coffman in Virginia; Todd Petty, Chris Shockey, and Gil Willis in West Virginia; Alan Heft in Maryland; Bill Ferris, Mike Heck, and Ken Undercoffer in Pennsylvania; Brian Cowden in New Jersey; Jeff Kirschman and Dennis Skarka, both of whom passed away in 2019, and Walter Ackerman in New York; James Belden, Joe Hovious, and Kirt Mayland in Connecticut; Michael Hopper (who passed away in 2018) and Mark Hattman in Massachusetts; Dan McKinley in Vermont; Steve Perry and Mark Prout in New Hampshire; Dave Boucher (who died unexpectedly in March 2016), Forrest Bonney, Merry Gallagher, Matt Libby, and Rick Young in Maine; and Cliff Randell (who died in a tragic floatplane accident in 2019), Ned Whittle, and Robin Reeves in Labrador. All these individuals were instrumental in providing background information on brook trout in specific states and streams, and they helped me catch fish in every state. They are also the folks who have done much of the heavy lifting in conservation and attend to the details in restoration work, usually out of the public eye. They are the guardians of home waters for all of us, and their efforts allow the angling public to continue to catch brook trout.

I also wish to acknowledge fellow anglers Mark Paulek, Tony Kuznik, Bo Ackerman, and Mason Overstreet. They helped make fishing trips a hell of a lot more fun and interesting. Mason, my former student and teaching assistant at the University of Alabama, and now an environmental attorney in Vermont, was incredibly generous with his time, energy, and enthusiasm in search of brook trout. I also want to thank the College of Arts and Sciences and New College at the University of Alabama for supporting my travels throughout the writing of this book. I am especially grateful to George F. Thompson, my de facto publisher, who has valued, supported, and developed this work from the start, and for that I am deeply appreciative. He took the manuscript to another level and found a home for it at the University of Georgia Press. And I am indebted to Mikki Soroczak at GFT Publishing for both her editorial and research assistance and the meticulous preparation of the manuscript. I am also indebted to the writers Rick Bass, Barbara Hurd, Alison Deming Hawthorne, Debra Marquart, and Helen Peppe, who provided invaluable comments on early drafts that were part of my thesis for the Stonecoast MFA Program in Creative Writing at the University of Southern Maine. I am grateful to friends and family for their enthusiasm and celebration of this book, including Martin Morthland, my parents, and my sister Jeanne. I am also most indebted to my ex-wife, Barbara Schlichtman, who encouraged and was supportive during all my travels. I am also deeply appreciative of Karen Talbot and her willingness to provide the art for this project. I knew the first time I saw one of her paintings of brook trout that her work would be a perfect complement to my words.

SEARCHING FOR HOME WATERS

Winslow Homer's *A Brook Trout* (1892), one of the most celebrated watercolors of the eastern brook trout (*Salvelinus fontinalis*).

The Eastern Brook Trout as Icon

THE BROOK TROUT is among the most iconic species for anglers in eastern North America. This lovely fish registers as a powerful symbol because of its beauty, imagery in art and literature, and role as an indicator species of the overall health of an environment. My affection for this species will, I hope, become crystal clear as you follow my journey from southern Appalachia to the mid-Atlantic states, New England, and Labrador.

My use of the word *lovely* gives away the reason I hold this small fish in such high regard. There isn't a more striking fish in the East. The sides of the brook trout are sprinkled with dots and dashes of red, orange, yellow, and the occasional blue. Its belly—especially in the autumn—turns bright orange, while its upper back is dark green or even black, providing stunning contrasts. The first time I held a brook trout, I thought that an artist had painted all the colors of a fall Appalachian forest on this one small, living palette.

Yet the camouflaged back of this fish makes it almost invisible in a stream. Not until we pull a brookie from the depths of its cool, dark pool can we see and truly appreciate these small masterpieces. Sometimes when I sit near a stream, I almost feel sorry for the hikers who quickly walk by, seemingly oblivious to the beauty of the creatures hiding among the cobbles on the stream's floor. The hikers' lack of attention in turn makes me think more deeply about what I might be missing in nature when I am distracted.

For centuries, the brook trout has inspired artists and writer alike. Its image has graced the canvases of America's best sporting artists, beginning in the nineteenth century with Winslow Homer's paintings of leaping brook trout. His celebrated watercolor of 1892, *A Brook Trout*, is one of the most recognizable images of fish in American art. It both depicts the species with accuracy and places the fish in an almost unimaginable aerial pose.

Homer's focus on the brook trout should not be particularly surprising given that, during the nineteenth century, it was the only salmonid in many eastern rivers, a prominent setting for his art. The brook trout remains a popular subject today among piscatorial artists like James Prosek, whose vivid watercolors of small brook trout are set among the streams of his native Connecticut; Joseph Tomelleri, whose scientifically detailed work has appeared in more than a thousand publications and whose client list (ranging from Bass Pro to Zebco) makes him perhaps the best-known fish illustrator today; Bob White, whose fish and fishing watercolors always inspire me to start planning my next fishing trip; and Karen Talbot, whose creations grace the pages of this work. These individuals are not only talented artists but also avid anglers who effectively depict the essence of fish and their landscapes.

The allure of brook trout and the landscapes where they are found are also reflected in an abundance of literature. Contemporary authors such as James Babb, Chris Camuto, John Gierach, Nick Lyons, Craig Nova, and W. D. Wetherell have written in great detail about the intersection of life and angling for brook trout. Their work has led me to contemplate the role of the brook trout in my own life. Other works focus more on biology and natural history, with Nick Karas's book *Brook Trout* serving as the standard-bearer for any and everything about brook trout. Before the mid–twentieth century, the connection between literature and brookies specifically is harder to trace, but trout in general and fly-fishing in particular provided fertile waters for a rich body of work. The great naturalist John Burroughs (notably in his essay "Speckled Trout") and the writers Theodore Gordon, Ernest Hemingway, Norman Maclean, and Henry David Thoreau all fished in brookie waters.

Beyond the creative inspiration provided by the brook trout is that its presence tells us a great deal about the health of the larger environment. The brook trout is an indicator species for streams, lakes, and watersheds. Because they are most plentiful in largely unspoiled conditions, their association with clean intact environments is another reason the brookie has developed a dedicated following among fly anglers. When I have a brook trout in my hand, I know the water in which I am standing is nearly pristine.

The writer Chris Camuto notes in *A Fly Fisherman's Blue Ridge*, "Wild trout are a sign the land is doing well." And, according to Gary Berti, a former coordinator of the Trout Unlimited Eastern Brook Trout Campaign, "brook trout are the canary in the coal mine when it comes to water quality. . . . Declining brook trout populations can provide an early warning that the health of an entire stream, lake or river is at risk."

The brook trout is also one of the most widespread species of trout, often described as the only native member of the Salmonidae family found in the eastern United States. This is technically incorrect, because the closely related Arctic char lives in a handful of isolated ponds in northern Maine, Atlantic salmon continue to swim in the waters of several rivers and lakes in Maine and many more in eastern Canada, and lake trout are also found in the deep lakes of New England. The brook trout, however, is the only native salmonid found south of New England, in highland areas such as the Allegheny and Blue Ridge Mountains.

The brook trout's native range roughly spans the spine of the Appalachian Mountains, from northern Georgia to northern Labrador and west to the Great Lakes and Hudson Bay regions. Within this vast geography exist such distinct populations as "coaster" brook trout in the Great Lakes; "salter," or sea-run, brook trout along the coast of New England; southern and northern strains in the Appalachian Mountains; and unique strains in some of Maine's ponds.

Habitat scale or opportunities correspond with the scale or size of the fish. Even so, it's strange that the small but mature six-inch brook trout I caught in foot-deep clear streams in South Carolina are the same species as the giant twenty-incher I caught in the deep-black rivers in Labrador. Geography matters.

Because of stocking efforts during the nineteenth and twentieth centuries, char and trout today are found outside their natural range and overlap with introduced species such as brown and rainbow trout. The brook trout is now widespread in western North America and has become an invasive pest, displacing many native cutthroat trout species. So while conservationists in the East work to protect brookies, those in the West seek to remove them. Their presence and its ramifications became clear to me while fishing with my wife in Montana's Lee Metcalf Wilderness Area, far outside the brook trout's natural range. For every native cutthroat we caught, we probably landed ten brook trout—hungry brookies introduced her to fly-fishing that day. Although that experience remains special, the memories are somewhat tempered by the knowledge that brook trout don't belong in Big Sky country. The experience would have been more pure had we caught only native cutthroats.

The environment in which the brook trout is found is also part of its past and present appeal. To fish for brook trout is often to fish in the last remote and rugged landscapes in the East, "fishscapes" not polluted by the stocking trucks that dump nonnative brown and rainbow trout

The native range (shaded area) of the eastern brook trout (*Salvelinus fontinalis*) in North America, despite significant changes in the environmental quality of habitat and waters of the eastern brookie since the early 1600s, when Europeans began to arrive. Map by Mary Lee Eggart.

in most of the East's accessible cold waterways. Part of the allure of fishing in general is the landscape, whether the jungle-like mangrove swamps of the tropics or the cold boulder-laden streams flowing from mountains. The brook trout's home environment—streams in the old hemlock-covered Appalachian Mountains and silent ponds in northern New England—are some of the most beautiful landscapes in the eastern United States. One inviting feature of these waters is that they do not suffer the same levels of fishing pressure as lower-elevation streams that are more accessible. It is common to spend a day completely alone while hiking up a brookie stream, even though you may be only a few hours from urban life along the east coast. During most trips to fish for brook trout, my only companions are brightly colored salamanders, birds in the treetops, and an occasional black bear. These streams are not home to large blandly colored, hatchery-reared fish.

Anglers who hike up to cold trickles appreciate the intangibles that nature has to offer: great views from the tops of mountains, the physical challenge of getting to rugged streams, and the chance to catch colorful native fish. When brookies strike at the end of my invisible leader line, the mountain has bestowed upon me a living gift.

Today brook trout in the eastern United States receive more conservation attention than they ever have, from growing numbers of local, state, and federal conservation and restoration projects. The simple yet understandable desire to catch big fish does not drive this interest. Brookies, in most of their range, are not and never will be big fish. I feel a sense of accomplishment when I catch an eight-inch fish in a headwater stream. The love of native fish and their landscapes is the motivation that largely drives the effort to improve and protect the brook trout's habitat. To me this is a healthier motivation than managing a species and ecosystem solely for producing trophies.

I am not naive about our past or current disruptions to landscapes of native trout. In almost all areas where brook trout are found, the ax and plow—if not the front-end loader—were close at hand at some point in the settlement of North America during the past three centuries, and nonnative trout have been stocked in every state. And before Europeans arrived, indigenous peoples also altered the landscape through trade, fire, and agriculture. Pristine forested landscapes are rare in eastern North America today, except for a handful of old-growth forests scattered throughout the remote mountains, such as in Maine's Baxter State Park. Nineteenth-century photos of the Green or Shenandoah Mountains show fields growing stumps and stone walls, not the mature forests we associate with these areas today. But wilderness can also be thought of in terms of personal meaning and experience and not just in relation to miles from roadways or thousand-year-old forests.

When I hike and fish a mountain stream, I sometimes like to imagine that its true headwater, or source, is in some remote wilderness well beyond my reach rather than my stopping point at the moment. Even if the headwaters are within reach, I often stop fishing before I reach them so that the stream retains some secrecy in my mind. Wilderness is a place where native species still dominate, where one can find peace, meaning, and some mystery in the natural landscape. These places are the mountaintops, gorges, and remote ponds where colorful brook trout still hover in cool dark waters. If wild and native brook trout are still out there, wilderness and clean pure water abide.

Home Waters and Symbols

FLY ANGLERS tend to subscribe to the notion that each of us has a home water. These are waters we know well and return to again and again; we also form deep emotional bonds with them. And part of our connection with home waters is the species we target. The idea of home waters also relates to the love of place, or *topophilia*. Coined in 1974 by the geographer Yi-Fu Tuan, *topophilia* describes the connections people make with specific places; the word literally means "a love of place." To have a home water is to have a love for a specific location or environment and the species that inhabit it. For fly anglers, a particular species often represents the love of place. Flats anglers have bonefish and permit, anglers in the West have steelhead and cutthroats, and we in the East have brookies.

The idea of a home water has been a prominent theme in fly-fishing literature. Chris Camuto, in *A Fly Fisherman's Blue Ridge* (2001), provides one of the best descriptions of a highly personal connection with a stream. He describes his deep ties to the North Fork of the Moormans River, a 14.3-mile tributary of the South Fork of the Rivana River in central Virginia. He seems to know every pool that holds fish, even individual fish found there, as well as how the river has changed through the seasons and years.

As I read Camuto's book, I envied his knowledge and long relationship with his Virginia landscape. As someone who has lived in many regions of the United States, sometimes for short periods, I have long wondered about the location of my home water, or topophilic fishing landscape. Is it the waters of my native Missouri? I was drawn to nature and fish from the time I took my first steps, partly because I spent my early childhood summers at the Missouri Botanical Garden's arboretum in Gray Summit, now known as the Shaw Nature Reserve, where my great-uncle served as the caretaker. My days in the reserve revolved around fishing and exploring the creeks, ponds, fields, and forests with various family members. Gray Summit, Missouri, may be an unfamiliar dot on the map to most readers, but the arboretum was the landscape where I formed my lifelong love for the outdoors. I retain pleasant memories of the arboretum—of catching stringers of black-backed bluegills with night crawlers, crawfish with traps baited with hot dogs, and the silky, slimy channel catfish using "stink bait" that smelled so bad I questioned whether we should actually eat those fish. It was my first topophilia.

On the other hand, I have wondered whether my home water is the Gulf of Mexico, where I spent so many Florida vacations catching speckled trout, whiting, and sheepshead when I was developing my angling skills in my adolescent years. There I tried to spend every hour fishing, no matter the wind or weather. My parents had to drag me away from the surf and bait bucket when it was time to go to dinner or the lightning cracked too close for their comfort. I still find turtle-grass flats and mangroves fascinating environments and still fish them regularly, now mainly in Belize. The slightly fishy smell of rotting vegetation, thick humidity, salt that stings my eyes after a long day on the water, graceful frigate birds soaring overhead on trade winds, and occasional schools of spooky bonefish and permit all conjure happy memories that continue to draw me back today.

As I began to spend serious time angling for brook trout for my brookie checklist and writing this book, I wondered whether I might now claim a new home water, one with which I have a far shorter history, and I continue to ask myself, Can someone shift their topophilia to a new angling landscape as they pursue new favorite fish? Or is one's topophilia tied to one's original home landscape: a single originating place that can never change? Since the turn of the new century, I have lived in and traveled to places from Maine to Hawaii, Central America, and Cuba. But that restlessness or wanderlust has also created a sense of placelessness that drives my question about my angling geography.

The shift in my fishing landscape and focus on brook trout began almost two decades ago during one particularly hot October in Baton Rouge, Louisiana, where my wife, Barbara, and I lived at the time. While I am comfortable in the heat and humidity of the larger Gulf coast, that summer heat carried on deep into the fall. The mosquitoes were still swarming, and evenings on the porch of our 1914 Craftsman bungalow were accompanied by streams of sweat rolling down my back. The evening air still stunk from the industrial humidity blowing in from the Mississippi River, flowing just a few miles from our house. It had been hot, truly hot, since early May. As Halloween approached, I was in the mood for chilly nights and some fall colors, for the air to be cleansed by cold fronts from the northwest. Barbara simply rolled her eyes at my discontent; she was and remains a much more tolerant person than I.

This desire for new weather fit my natural tendency toward wanderlust. My reading habits feed my desire to travel and explore. I read books about places I think will make me happy or where I want to be at a particular time, mostly nonfiction such as Peter Matthiessen's *The Snow Leopard* (1978) but also fiction such as Jack Kerouac's *The Dharma Bums* (1958). I am a geographer, so I like authors who take me on journeys, real and imagined. The weather also seems to have an inordinate influence on my reading list. When I am cold or have been away from the Gulf coast for too long, I page through my dog-eared copy of *Fishing the Flats* (1983) by Mark Sosin and Lefty Kreh or reread Ernest Hemingway's *Islands in the Stream* (1970). Hemingway's descriptions of the various blues and greens of the Gulf Stream, icy daiquiris, and conch salad deliver me to Bimini. I had complained enough about the heat that a sympathetic and transplanted Yankee friend gave me his copy of *Vermont River* (1993) by W. D. Wetherell. This marvelous book, with its seasonal descriptions of his unnamed home river and of trout fishing, created a deep desire in me to stand in cold rushing water while casting dry flies to rising trout. I wanted to see his home waters and fish for his favorite fish.

I was not teaching at Louisiana State University that fall, so I rashly thought about cashing in my frequent flier miles for a quick trip to Vermont. Wetherell had placed me in a cooler landscape. Unlike Hemingway's tropical haunts, this one had frigid rivers, colorful trout, and dark spruce forests. Since it was late October, I knew I was pushing into early winter in the Green Mountains, so I had to act fast or wait many months to feel the same motivating forces of heat and bugs. Barbara barely had the time to ask "Where?" before I had booked my trip, packed a bag, and was out the door. Since she had heard me complain all fall about the heat and seen the trout books on the nightstand, she was not surprised when I announced my retreat and probably happy to have me out of the house for a few days.

I had fished for trout only once, at a state park in my native Missouri when I was a boy. So that fall's reading was an introduction to wild brook trout and mountain streams. My childhood experience with trout entailed standing shoulder to shoulder with other shivering people on an opening day in late winter as I tried to catch a few hatchery-raised rainbow trout and avoid getting snagged by the treble hooks on my neighbor's miniature Lazy Ike. I loved to fish, but even at a young age I associated fishing with summer heat and fewer people. So I felt strangely out of place. I found nothing inspiring about being crammed together with five hundred strangers on a slate-gray, pre-dawn morning at Meramec Spring, a Missouri state trout park a few hours from my childhood home in St. Louis.

My very first cast on my very first opening day might have contributed to my lack of interest in that kind of trout fishing. A moment after the opening whistle sounded, on my first cast ever, my brown woolly bugger spinner became entangled with the line of the angler fishing directly across the stream. Both of us felt resistance on the end of our line, both set our hooks, and we ended up with a monofilament tightrope, one high and tense enough that a small flock of swallows could have roosted on it. I was momentarily panicked. My cohorts were pulling in dazed and frozen ten-inch trout, seemingly on every cast. I feared none would be left when I got my act together. The gentleman with whom I was entangled was nice enough to cut his line instead of mine. I have always appreciated that kindness.

And while I did catch a handful of bland-colored stocked fish, that day witnessed both the beginning and end of my boyhood trout-fishing career. The frenzied pace, large crowds, and stocked fish did not fit my image of trout fishing. That place could never serve as home water, and the fish could never be special. I grew up watching Curt Goudy on *American Sportsman* as he caught native cutthroat trout in his native Wyoming and the American West. In my twelve-year-old mind, I associated trout fishing with few people, wild rivers, and native fish. I was a trout snob and didn't know it. Also, my father was not a trout fisher. He preferred bass, crappie, and catfish, which meant I preferred bass, crappie, and catfish. So my early trout experience was skewed.

While trout were not part of my fishing world until my early thirties, I always considered the brook trout, with its gold and reddish markings and its camouflaged green and black back, to be one of the most beautiful of freshwater fish. Although I was unfamiliar with its haunts, the species was special for me. At a young age, I subscribed to *Field and Stream* and *Outdoor Life*, so I occasionally saw photos and read articles about brook trout. Since I was born and raised in Missouri, I also found brookies appealing because I associated them with remote, mysterious mountain streams in deep hemlock forests well off the beaten path and far from the muddy rivers around my home. Pictures in the magazines, of alpine lakes and rivers with spruce and hemlock forests in the background, made the brook trout seem exotic to a midwestern kid. The species seemed "out there," far from any streams to which I had access.

I had only cruised through Vermont in 1987 on my way back to Missouri after a college summer spent in Maine. But as I read Wetherell's book during that hot fall in Baton Rouge, his descriptions of fall colors and country scenes awoke memories of that first trip through the Green Mountains. The book gave me a sense of confidence that I was heading back to familiar territory and that Wetherell's story would soon be my story. The book allowed me to see myself fishing and hiking up those spruce- and maple-lined streams while methodically casting to every fishy-looking pool.

When I stepped outside the airport terminal in Manchester that late October night, my breath was visible in the air. The next morning, heavy frosts covered roadside fields. Frost on cornfield stubble twinkled brilliantly in the dawn sun, and steam rose from barn roofs. Based on the still slightly colorful piles of leaves in yards, most of the trees had shed their leaves only recently, but enough color remained to reveal the beauty of New England in late fall. Halloween pumpkins on front porches of old farmhouses added to the seasonal appeal. And the barren, almost skeletal, trees somehow gave the landscape an even more mysterious look in my mind's eye.

Before I left home, I had found a guide on the internet and hired him to help introduce me to the mountain trout fishing that Wetherell describes. I met the guide near Killington at 8:00 a.m. and drove up VT 100, a picturesque road that runs north-south along the spine of the Green Mountains. We entered the Green Mountain National Forest near Pittsfield and bounced up a steep rutted U.S. Forest Service road. I rolled down the truck's window and breathed in the cold balsam-scented air. I had layered all my winterish clothing, which did not amount to much, given that I lived in Louisiana. We finally stopped next to a small stream peppered with mossy boulders.

As most small-stream trout anglers know, our fishing consisted of quietly climbing up the stream, slowly sneaking into position, and casting to each small pool. I never thought I would fish for trout on my hands and knees, although this was necessary in some areas, because the stream was small enough that fish in the miniature pools would detect anyone or anything approaching. I learned quickly through trial and error. Much to my chagrin, my sometimes loud stumbling approaches sent small trout darting before I ever cast my fly.

I was several miles from the nearest paved road—no crowds, no cars, no human sounds, just a late fall forest and rushing water. I had missed the prime fishing and leaf-viewing season and had arrived just before hunting season, so my guide and I had the landscape all to ourselves. As I look back with a present-day appreciation of how much that first experience of catching brook trout affected me, some lines from *Moby-Dick* (1851) come to mind: "Take almost any path you please, and ten to one it carries you down in a dale, and leaves you there by a pool in the stream. There is magic in it." Indeed, I found magic in those Green Mountain streams.

After that first day, I said goodbye to my guide. I was his last client of the season, and his impatient casting instructions told me that he was ready to end his fishing year and move on to deer and grouse season. No matter. I thought I understood the basics, so I wanted to set out on my own to fish and explore the central Green Mountains. Small stream fishing doesn't require long picturesque casts. During the next several days, with a borrowed fly rod and a handful of flies, I crisscrossed the central Green Mountains alone, climbing streams and catching small native fish. Anglers take great satisfaction in exploring, finding, and then actually catching fish in terra incognita. Using the *Vermont Atlas & Gazetteer*, I found accessible streams in the national forest and followed them upslope, trying any pool that appeared big enough to hold fish.

I did not have a wide assortment of flies back then, and those consisted of the basics such as elk hair caddis, Adams, a few oversized stimulators, and tiny copper johns that I purchased at a Vermont country store. I tried nymphs and dry flies; some worked, some did not. I missed a lot of small fish, I left many flies in the trees, but overall the brook trout and occasional rainbows did not seem too discriminating, and it was apparent even to me, then an inexperienced trout angler, that those small fish did not get a lot of pressure. Brook trout in such conditions are confidence builders for the novice fly angler, so I was pleased with myself after landing fish in a new landscape. I would have been happy with any fish, but the beauty of fall brook trout made the experience even more memorable. What a way to escape the southern heat.

I do not fully understand how our mind and soul develop affection or love for a place or why specific species become important to us. Is it the smells, colors, physical features, plants, birds, fish, or all these that figure into the development of our topophilia, or home water? Whatever the exact process, I developed a love for those streams and their fish high in the mountains. It might be corny, but when I dropped into the Green Mountains that first time, I felt an instant bond with that larger landscape of brook trout and the fish that I and many others designate as our favorite species. I felt like I had found my home water. When a trout, even a five-inch fish, rises from the bottom of a black pool and violently takes a fly, that moment is electric. The stream seems more alive when fish are drawn out of their hiding places.

Even with so little trout-angling experience, I knew intuitively that brook trout were somehow nobler than bass and catfish. My bass friends in Alabama would surely disagree with me, but then every opinion is weighted with bias based on experience. The trouts' bright spawning colors in the fall, their eagerness to attack a fly, and their cool quiet home instantly drew me in. The trout I caught were smaller than Curt Gowdy's, but they were native fish in a wild stream. Similarly, their landscape seemed more environmentally intact than the muddy creeks of my childhood and present-day home. I saw no homes, no other people, no discarded beer cans, and heard no sounds other than rushing water and the occasional white-breasted nuthatch and small flocks of chickadees in the trees above me. I was also drawn in by the ethics of fly angling. Many, if not most, fly-fishers practice catch-and-release fishing. While I have no problem eating stocked fish, I always release native trout. The fish I caught during that first trip were no exception. As the great cold-water conservationist and innovative fly angler Lee Wulff wrote, "Gamefish are far too valuable to be caught only once."

Also, if brook trout are present, I am reasonably confident the water quality is good, in contrast to the creeks and rivers of my youth. Many of those rivers now welcome the public to their banks with signs warning not to consume too many or any fish at all. In one infamous case, Times Beach, a small town on the Meramec River just south of St. Louis, helped push Congress to pass the Superfund law in 1980 and was designated a federal Superfund site three years later because the soil was found to contain high amounts of dioxin. The U.S. Environmental Protection Agency bought out the entire town of 2,220 people with Superfund money; the site was completely abandoned by 1985 because of contamination. I camped and fished upstream from that river town when I was a young boy.

That type of environmental degradation eroded the relationship I had with my boyhood home waters and their fish. Instead of being awed by the beauty of nature, I began to see industrial pollution. And around St. Louis, with its history as an industrial center, one does not have to look far to see such damage. A drive along the riverfront in St. Louis is a drive through an environmental wasteland, with decaying and rusting factories, many of which are now abandoned. This description sadly applies to many cities whose industrial cores have been decaying for decades, leaving behind toxic legacies.

I was diagnosed with thyroid cancer when I was thirty-three. As I was being treated, doctors asked, "Have you ever been exposed to radiation? Were you in Europe when Chernobyl exploded?" I always shrugged, but I have thought since that perhaps I was exposed to some industrial pollution while wading, fishing, and playing in the many slimy creeks around my Missouri home. God knows I spent enough time in them, ruining one pair of sneakers after another in creeks whose water often had an oily sheen. I will never know what pollutants I might have been exposed to, but I will never stop wondering. When your body fails you at a young age with no clear explanation, it is natural to seek out explanations. Today I do not look at polluted waterways in the same naive way I once did. My sons do not play in streams with oily rainbow sheens like I did.

For years after my surgery, every ache and pain sent me into a panic. I obsessed over whether the cancer had returned. In that period, I realized, too, that I often try to live life too quickly, moving, changing jobs, obsessively writing, always looking for the next adventure rather than being present in the moment. My illness contributed to my sense that my personal geography is chaotic. Writing this book about my travels to fish for brook trout from Georgia to Labrador is one way I have taught myself to slow down and to be mindful, to be present in a specific place and a specific activity. In many ways, fly-fishing a mountain stream, usually alone, has been my form of meditation. When my obsession with brook trout began fifteen years ago, it was

I'm casting into a headwater stream in search of native eastern brook trout in Great Smoky Mountains National Park in North Carolina. Photograph by Robin Boggs, 2020.

really the first time the long shadow of my disease never entered my mind. Thomas McGuane wrote in his brilliant collection of fishing essays, *The Longest Silence* (1999), "I had in my own heart the usual modicum of loneliness, annoyance, and desire for revenge, but it never seemed to make it to the river." When I am fly-fishing, I feel no mysterious aches, pains, or creeping fear of a cancer; they do not make it to the river.

In fact, for me, fly-fishing is the only activity during which my mind becomes totally clear, my focus allowing for no distractions. And, again, no mysterious aches and pains darken my mind. I liken this clarity to the tenets of a Zen Buddhist practice I developed after I graduated from college. Fly angling and Zen are interwoven, because both require my mind to focus on the *here* and *now*. Peter Matthiessen, in his *Nine-Headed Dragon River* (1986), describes this mindfulness as "rest in the present." But this focus and rest do not mean I am thinking about only the immediate fish or cast; this focus brings all parts of the physical and intellectual being together to land a fly so softly that nary a ripple appears on the water's surface. As I wade up or down a stream, casting to productive-looking spots, my mind is completely congruent with my physical actions. Without any conscious physical actions to direct it, my fly often lands in the spot my mind chooses. If I do have a frustrating moment or outing on a stream, the beauty and serenity that surround me usually bring me back to the present, even if the fish aren't cooperating.

Looking back on all these experiences and thoughts, I can see I was emotionally primed for a new landscape when I first fished in Vermont and encountered brookie fishing. While we lived in Baton Rouge during the 1990s, Barbara and I had a rueful routine about the reaction of state officials to the most recent chemical spill or toxic gas release.: we would solemnly repeat to one another, "The public was never in danger," because that's what the press releases inevitably claimed. Sadly, several of our Louisiana friends have since passed away after contracting various cancers. I love many things about the South—including its food, music, literature, football, and such natural environments as bottomland forests and coastal marshes—but its environmental record is a disaster. Mining, the petrochemicals industry, and industrial agriculture run roughshod over environmental regulations in much of the Deep South and, sadly, in much of the United States. For example, in my current home state of Alabama, the various river keeper organizations have to step in and legally force or shame the state to respond to spills or other environmental violations because the very industries that threaten the environment passively control state agencies. The mingling effects of my past illness, nearby industrial pollution, and debilitating late fall heat pushed me to find a new fishing landscape, one away from a chemical corridor along the Mississippi River known as Cancer Alley. Central Vermont satisfied that need.

I also found a geographical connection to brook trout, because their range extends from northern Georgia to northern Maine and Labrador. The brook trout is the only native salmonid found south of New England, in highland areas such as the Alleghany, Appalachian, and Blue Ridge Mountains. Their range has come to represent my range.

I have spent much of my adult life living in and traveling between the Deep South and New England. My job as a geographer has taken me to Baton Rouge and then

Tuscaloosa, with many stops in between, but since 2011 my family has remained in Maine. We wanted to raise our sons in Maine, a place we knew and valued for many reasons, some of which are described herein. But the distance and my commute took a toll on our marriage, which ended in 2017.

In Maine, my kids assume trout live in the rivers, the water is clean, and the surrounding slopes are forested. While I remind them that was not always the case—that many, if not most, of Maine's rivers such as the Androscoggin were highly polluted until relatively recently—I find it reassuring that they assume a high level of environmental quality in their home place. And I hope their youthful assumptions make them dedicated environmental stewards.

Sometimes I am envious of Truman and Frederick, now in their twenties, because their introduction to trout streams and fly angling took place more than a dozen years ago in the eastern corner of the White Mountain National Forest in Maine. On a warm and glorious July day with a bluebird-colored sky, we had a stream all to ourselves. We shared a three-weight fly rod, stripping weighted wooly buggers through deep pools. We caught more than a dozen native brookies and wild rainbows. Late in the afternoon, I sat atop a smooth van-sized boulder warmed by the sun and watched them mangle casts, laugh, and fish with a carefree sense about them. They know nothing of crowded stream banks, tangled lines, and stocked fish. They have no idea how lucky they are.

So, as I traveled the spine of the Appalachian Mountains every May and August from Alabama to Maine and back, I was comforted, because I was in the home range of the brook trout's landscape for almost the entire trip. Thus, I was in my newly adopted home water the entire time. Even when I fly north and south today on shorter trips, from my window seat I can sometimes identify the specific landscape features far below and even point to areas I know are home to brook trout. I know that landscape on the ground and from thirty thousand feet. Sadly, life takes turns that are unexpected. Because My marriage didn't survive the geographic distance between the South and New England, my time on brookie streams in Maine has greatly diminished. Although I spend much less time in Maine today, brook trout landscapes remain my connection to, and a powerful metaphor for, my family life. The association is deeply nostalgic, given what I had and lost and the memories made during that part of my life, especially when my children were young. Brook trout landscapes remain my topophilic home waters.

A Gallery of Watercolors

From the Land of the Eastern Brook Trout

BY KAREN TALBOT

KAREN TALBOT is an award-winning illustrator and celebrated artist from Maine who believes in advancing environmental awareness and conservation through art. For many years, she has masterfully captured the essence of birds, fish, and other wildlife and botanicals she encounters while exploring the world with sketchbook in hand. In the tradition of Winslow Homer and John James Audubon, her renderings in a variety of media combine the details of a scientific illustration with the aesthetic beauty of what she draws and paints. Talbot also selflessly gives back a percentage of her art sales to help conserve the wildlands and waters that inspire her.

This gallery was curated by George F. Thompson, who suggested that Karen collaborate with Mike, who would suggest to her which flies, fish, birds, and amphibians he associates with the fourteen U.S. states and one Canadian province where he roamed, fished, and sought home waters. All of Karen's fifteen watercolors were thus made especially for this book and are presented from north (Labrador) to south (Georgia) with the title of each watercolor in italics, followed by a short caption by Mike. In this way, Mike's textual journey from Georgia to Labrador is reversed by the directional sequence of Karen's art. Each watercolor is used with the permission of the artist.

Copyright © 2021 Karen Talbot Art. All rights reserved.

This Labrador brookie was the last and biggest fish I caught on my journey.

The Maine blueback (*Salvelinus alpinus oquassa*) is a rare relative of the brook trout.

The barred owl (*Strix varia*) was a camping companion in Vermont.

The salter, or sea-run brook trout, is now the focus of many stream restoration projects.

This native Connecticut brookie was a real beauty.

March brown mayfly

Tied by Dennis Skarka

#10 dry fly hook
8.0 orange thread
2 wood duck feathers
piece of spade feather
fox fur (tan w/ a little yellow mixed in)
1 grizzly (rooster saddle) feather
1 ginger feather

Anglers throughout the eastern United States use
the Catskill March Brown, a popular trout fly.

Although not native to the eastern United States, the spring-creek rainbow is nonetheless a beautiful fish.

Pheucticus ludovicianus

"Rose-breasted grosbeak"

Stocky, medium-sized songbird
V. lg., triangular bill
Very little sexual dimorphism in size
Adult ♂
   black + white w/ brilliant red chevron
   from black throat to middle of breast

Chevron
924 922 925 924 937
    926 930

1100

997 1080 1085

no black ink!
watercolor base

747 for blending head/back

head/back
995 996 1007 1100 901

937 932 994 1050 1052

931 1078 930 1009

1008 1065 1067

1101 1063 1056

919 1004 987 1061

1088 1005

1098

152 46919

902 1076

1099

Long-distance migration — individuals
typically fly across Gulf of
Mexico in a single night!

Known for exceptionally beautiful song
like a robin w/ opera training

Artist study of the rose-breasted grosbeak. This bird (*Pheucticus ludovicianus*) is a summer visitor
in the Appalachians at elevations of 3,500 to 5,000 feet.

This West Virginia red eft (*Notophthalmus viridescens*) is a common newt in Appalachia.

"pure" (song)   "spotted" (breast)

Catharus guttatus   "Hermit thrush"

059 070 010 111 1280 035

168

Differ based on region:

Pacific states: smaller, thinner billed, w/ dusky brown upper body + greyish flanks

Mtn West: bigger, thin-billed + greyer overall w/ larger spots + less rufous

Eastern: medium-sized, thick billed; most colorful – buff on flanks + under tail

All have rich brown upper + smudged spots on breast, round head, long tail;
  warm reddish tail

thin, pale eye ring, distinct spots on throat, pink legs

No obvious sexual dimorphism

Primary feathers do not extend beyond secondaries (migrate short distances)

1034
140
1012
940

948   1074
947   1076
946   1056
1082 1052
941
943
5
937
1094
1028
1098
1084
1083 Walt Whitman!
1085

beak too thick?

Use exacto for
mottling?
(or th nim grey)

erase
pencil
lines
before
painting

keep
reflection
warm
(no blue)

pink, skinny legs!

150
1001
927
1020

942
917
1003
1002
918
1033
921
1032
988

Artist study of a hermit thrush.
The *Catharus guttatus* was a regular
companion throughout my travels.

I saw large flocks of this common nighthawk (*Chordeiles minor*) while in Asheville.

A Tennessee brookie, a mountain jewel.

Setophaga fusca "dark"

"Blackburnian warbler" – named (by Müler) in honor of Anna Blackburne – English naturalist
1776

black ink
blue CP

All adults: brilliant orange throat
- orange-yellow eyebrow
- sm. black face mask
- broad white wingbars

5188   180   127   146   110
107   PM-19   113   cd 140
CdA 280   113+107   060
247   111   cd-010

Breeding ♂
- throat + upper breast deep orange
- sides of neck, eyestripe, line on forecrown + eye arc yellow-orange
- face patch, crown + back black

741   916   917   1034   1003   1076   1067
948   1025   1027   1024   1063

light

Hangs out in upper pt. of canopy – breeds in mature coniferous forest (or mixed con/decid)
Only flocks during migration – otherwise territorial/solitary
           although forages w/ flocks of chickadees, etc., after fledge

Artist study of a Blackburnian warbler (*Setophaga fusca*). I saw these birds in forest canopies in South Carolina.

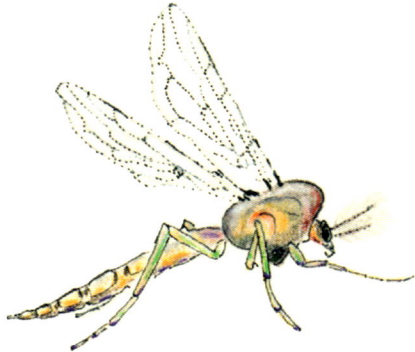

A midge. Midges are any small flies and common aquatic insects; anglers use imitation flies.

PART I
Southern Appalachia

CHAPTER 1

Georgia

THE SOUTHERN FINGERS of the Blue Ridge Mountains fan out into the northern quarter of Georgia, creating a landscape with suitable conditions for trout and other Pleistocene castaways. Although brook trout inhabit only 150 miles of Georgia's streams today, a minuscule geography compared with their habitat in states to the north, northern Georgia still has the look and feel of trout country. And while brookies' total habitat today may seem minimal, it's pretty impressive given the southern latitude of northern Georgia. Some of these "trouty" landscape features include the Appalachian Trail, which begins on Springer Mountain, as well as dozens of freestone streams and waterfalls in the Chattahoochee National Forest. That landscape also includes Blue Ridge Fly Fishing, a first-rate fly shop with operations in both Blue Ridge and Helen. The small town of Blue Ridge, located about ninety minutes north of Atlanta, is also home to Oyster Bamboo Fly Rods, which offers classes in making fly rods and sells exquisite custom-made bamboo fly rods that cost a small fortune. I participated in one of the weeklong classes and came away with a beautiful honey-colored, four-weight bamboo fly rod. I have dubbed it my "natives" fly rod, because I use it whenever my angling is focused on native trout such as brookies and cutthroats. It feels old school to catch native fish with a fly rod I made.

In early spring, before my semester ended at the University of Alabama, I traveled to Hiawassee, a small tourist town on the edge of the Chattahoochee National Forest, to meet with Leon Brotherton, a biologist with the Georgia Department of Natural Resources who directs restoration work for brook trout habitat. As I sat on the banks of the large and many-fingered Chatuge Reservoir (created by the Chatuge Dam), it was hard not to notice Hiawassee's recent building boom. Access to the water and Chattahoochee National Forest plus the area's mild winters have created a natural draw for vacationers and retirees alike. And like many other areas of the southern Appalachian Mountains, the landscape around Hiawassee is undergoing large-scale environmental changes because of development. At the time of my visit, vacation homes were sprouting up on the hillsides like acorns on the spring forest floor. Steep hillsides, too low to call mountains but steep nonetheless, were carved with switchback roads leading to summits from which the occupants of a single house could enjoy the view, one increasingly filled with human creations. This was the beginning of a similar pattern of growth I would see not only in Georgia but also in western North Carolina and eastern Tennessee. Everyone wants a piece of the mountains and, they hope, a view.

Not all the results of growth have been negative. Members of Trout Unlimited (TU) from northern Georgia and the greater Atlanta area provide a lot of the labor, money, and political influence that contributes to brook trout restoration projects and general cold-water conservation efforts in Georgia. So population growth has been beneficial from that standpoint. More people in the region want and care about trout. Still, as I sat in a lakeside parking lot waiting for a day in the field with Brotherton, I wondered whether the region's reservoirs could continue to satisfy the growth of both northern Georgia and Atlanta's ever-expanding suburbs. As I learned throughout my travels, water is an issue not only in the West but increasingly in the humid East.

Brotherton, a mountain man, trapper, and biologist with a deep Georgia twang, is from the Hiawassee area. He grew up fishing for brookies in some of the very streams he works on today. The forests and streams he monitors are his home waters. While state TU chapters provide much of the restoration labor, Brotherton monitors and patrols streams in this section of the Chattahoochee National Forest long after the volunteers have gone home. He is fortunate that, in addition to TU volunteers, he also has a team of paid interns during the summer months. This labor pool has allowed Georgia to work on improving brookie habitat, stream by stream, structure by structure, resulting in one of the most organized and intensive brook trout conservation programs in the eastern United States.

As in other southern states, most of Georgia's brookie streams are located on public lands—mainly national forests—so Brotherton does not have to convince landowners he isn't a threat to their livelihoods or property. We headed high into the Chattahoochee National Forest in his state-issued truck to look at some of the stream structures that volunteers and interns had built the previous year. We drove up and around a remote U.S. Forest Service road that had not seen another truck for most of the past winter, based on the amount of downed tree limbs blocking the road. Several times we had to get out of the truck and cut our way through fallen timber with Brotherton's chainsaw. With no cars and no people, when we weren't using the chainsaw, we heard few sounds other than the occasional crow and blue jay, and the area felt far more remote than it actually is, given the recent growth. We eventually found our destination: a level clearing in the forest that had been a homesite before the area was designated a national forest in 1920.

Although Georgia is at the southernmost reach of alpine environments in the eastern United States. The high elevations disguise the relatively short distance to a subtropical coast. A few ancient apple trees—gnarled, twisted, and hollow, a telltale sign of an old farm—sprout early spring buds on the tips of their meager branches, adding to the forest a bit of cultural mystery. Who planted those trees? Surely, whoever it was had fished for brook trout, locally known as specs, in the nearby stream. I like finding remnants of previous occupants, whether an arrowhead or apple tree, because they force me to acknowledge that someone else walked, farmed, loved, and died in a spot that I usually pass through with few thoughts other than to find a stream. When I am hiking through a forest and see a stone wall, I often wonder if the folks who took the time to build the structure also fished for brookies. Their concept of home water was probably far more utilitarian than mine.

Next to the clearing, a creek weaves its way through the regenerating forest. What probably was once a much larger clearing had become a tangle of brambles and small scrubby trees, with a few hardwoods trying to reach the sky and begin their eventual domination. The stream was a small headwater variety, not more than eight feet across at best. Before the volunteers had tackled it, it had been dominated by shallow riffles, largely devoid of any significant cover or pools. According to Brotherton, "there were hardly any brook trout in this stream before we got here, and those that were in the stream didn't last very long, because the habitat was so poor."

As forests were cleared during the nineteenth century and, in some locations, recently, valuable standing trees were removed along with trees and large branches that might topple over into and across streams and create habitat for trout, aquatic insects, salamanders, and many other creatures. Young regenerating forests do not contain much potential structural material, because young trees normally do not topple over. And streams barren of such structure are neither diverse nor ecologically productive. This connection between forests and stream habitat is lost on many people. Old forests feed streams. Remove the old

trees and more than just terrestrial ecosystems change; stream environments degrade over time as well.

Given the intensity of logging and farming, with many areas of the east coast's interior mountains completely cleared, it is not surprising that brook trout habitat today is poor in so many headwater streams. Under Brotherton's direction, volunteers and interns carefully placed sections of cut trees—usually trunks—in specific locations to create better habitat for brookies. The workers made up for what nature could not provide, given the age of the forests. Brotherton may be a biologist, but he has the intuition of an engineer and the instincts of a landscape architect, manipulating the stream at just the right spot to create the best possible habitat. Some of his creations include small falls, precisely placed log dams that created deeper pools, and chutes to channel the streams' energy to deepen a run. As we walked and talked, I turned over several stones in the recently created pools and found scores of insects clinging to the slick rocks, including several healthy-looking stone flies, common denizens of good trout water.

Georgia will likely never be able to grow brookies of fifteen or more inches in its headwater streams, for the streams in which brook trout are found are too small and often lack the sheer mass of food necessary for fish to grow large. But with better habitat in its headwater streams, Georgia will at least be able to hold on to its only native trout species and occasionally grow a nice size fish. And by creating deeper and bigger pools, Brotherton and his TU work crews also reduced the chances of brookies' disappearing during dry and especially hot summers. He told me, "Look, when we get those really hot and dry summers, like we had the last few years, trout really struggle to hang

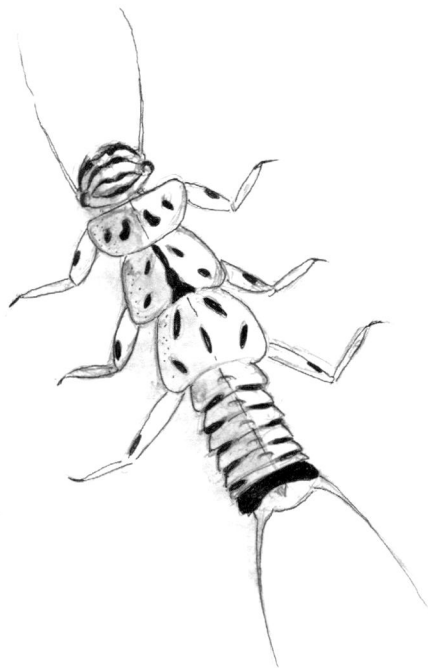

Stone fly nymph
(*Acroneuria abnormis*).
Drawing by
Frederick Steinberg.

on in some of these little streams. But these deep pools will make sure there are at least a few refuges, even in the hottest months."

Creating the best possible habitat in national forests is also important because in states such as Georgia and South Carolina brookie streams outside protected areas are few and far between. As I drove around Chattahoochee National Forest on my first and subsequent visits, I understood the desire for a home on a hilltop near a trout stream. The view of the hazy mountains, recreational opportunities, and proximity to wildlife are all motivating forces, but, I wondered, where will it end? As an outsider and tourist, I openly acknowledge my role in the development of previously rural spaces. I have stayed in the

same developments I criticize. I once brought a group of my University of Alabama students to Blue Ridge while teaching my class on the art and science of fly-fishing. We fished the Toccoa (Ocoee) River, caught stocked rainbows, and stayed in one of the big new vacation homes. Again, we all want a piece of the mountains.

The influx of people also means private stocking. No hard data exist in terms of who is stocking what and where, but it indeed is common in northern Georgia for vacation rentals to include access to privately stocked sections of streams. Although this stocking is not taking place in headwater streams, because most of those are found in national forests, rainbows and browns move around, often finding their way far from the original entry point or spawning grounds. So the efforts of these amateur fish managers could further constrict brookies in this region. I fished on one such private stream outside Blue Ridge and caught several rainbows longer than eighteen inches—the largest was twenty-two inches. Small brook trout won't be recolonizing that stream anytime soon. Still, while I lamented the changes brought about by stocking, I certainly enjoyed catching the biggest rainbows of my life. A salmon-sized trout on the end of the line provides an intense shot of adrenaline, a primordial thrill, and almost made me forget about its role in the decline of native brook trout in Georgia and elsewhere.

The day after our meeting and tour, I took Brotherton's suggestion and fished Stover Creek, high in the Chattahoochee National Forest, a 750,000-acre wilderness that includes 5,000 acres of old-growth forest. As I drove up the steep and narrow U.S. Forest Service road and then hiked in to fish, I quickly forgot about the encroachment of the modern world downslope. I was home again on a quiet stream with not another person in sight.

Stover Creek, which runs adjacent to the Appalachian Trail, is a small headwater stream that has been a focal project for Brotherton and his work crews since 2012. Stover is a feeder stream for Noontootla Creek, which, farther downslope, is home to those huge rainbows I caught. Thus, like so many streams in the East, Stover represents a last stand for brookies in this system. It is near the top of the watershed, so brook trout have no higher place they can swim to. Although the beautiful little stream was surrounded by mature highland forest and thick damp rhododendron, it lacked ideal brookie habitat before the volunteers swooped in and installed in-stream structures. This had become a classic brookie stream, with tight, crowded, greedy limbs that seemed to reach out to grab my flies. Frustrating, yes, but I wanted to meet a Georgia brookie in the worst way after a steady diet of stocked fish downstream, so I considered a few lost flies a necessary sacrifice.

On that cool spring afternoon, pools held plenty of opportunities. I made a few casts to each, caught a small brook trout, and moved on. A pileated woodpecker followed my progress along the stream, scolding me with its high-pitched call of *wuk, wuk, wuk, wuk*, a good sign that I was in a mature forest. After catching several fish, I realized how mechanical the fishing was. I was distracted, still thinking about all the switchbacks up the sides of once-forested mountains. I hadn't let go of the sounds of hammers and saws, mating calls of the world encroaching on the national forest, and my own role in that development. I was rarely distracted on a trout stream, so it was a new, uncomfortable feeling. I hadn't reached any sort of Zen-like equilibrium between mental and physical, of being present in the moment. I stopped and sat down a few yards from one especially promising-looking pool. After a few minutes, a small brookie appeared, feeding on tiny midges that swarmed in the slack water. Pop, pop, pop, the brookie feasted, poking its head out of the water ever so slightly. After watching that fish feed and studying the stream, I broke down my fly rod and headed up the trail for a hike and to eventually make camp. All was well on that stream. I looked forward to night sounds next to the brookie stream on a moonless night.

The experience grounded me again, because it placed me in what felt like endless wilderness on both a large and a small scale. From the trail I could have walked and fished all the way to Maine. For a few moments, I fantasized about that option: a mendicant lifestyle as I fished my way north. It was a tantalizing idea, to fish and walk all the way home, and it reignited my sense of wonder. Wonder about the large scale—the Appalachian Trail— and also about the small scale of that single pool and its small but perfect fish. That pool, too, was wilderness on a small scale, no matter the changes taking place outside the national forest. As I stood on the trail looking north, it was easy to breathe again, to breathe deeply and start walking. Just as those thoughts were rolling through my mind, the harsh call of a pileated woodpecker, zooming through the canopy, reminded me to get back to the fish.

CHAPTER 2

South Carolina

AS I LEFT Interstate 85 after a six-hour drive from my academic home in Tuscaloosa, where I am a professor of geography and teach at the New College at the University of Alabama, I turned north toward Seneca, South Carolina. I saw nothing that told me I was anywhere near trout water. The red earth looked like the same iron-rich oxisols I had seen earlier that day in Alabama and Georgia. The sluggish coffee-brown streams I passed looked like they would more likely be home to gar than trout. As I rolled into a parched roadside gravel lot to stop and stretch my legs, the early summer heat and humidity that had been enveloping me since Tuscaloosa remained a tormenting force. Standing there sweating, with early-season katydids buzzing loudly above the passing traffic, I felt like I was a million miles from cold water.

Of all the eastern U.S. states I have fished, and I have fished all the mountainous states, I was most skeptical about South Carolina's brook trout potential as I made the long drive across the sunbaked southern landscape. Its physical geography seems to fit in more with the larger coastal plain—hot, humid, and swampy. And its political battles about Confederate symbols mean its cultural geography seems closer to the former plantation South than highland Appalachia. But when I entered the northwestern corner of the state, my outlook shifted dramatically. Instead of Lowcountry live oaks draped in Spanish moss, I found highland forests of hemlock, white pine, and tangled groves of rhododendron and mountain laurel.

The Blue Ridge Mountains, part of the larger Appalachian Mountain system that roughly stretches all the way to Canada, reach their southern terminus in northwestern South Carolina and nearby Georgia. And while the highlands make up only a sliver of South Carolina's geography, this sliver has provided species such as brook trout with a niche in the Deep South. When I pulled into Seneca and checked into my cinderblock, 1970s-style hotel, I saw a Walmart, Wendy's, and Zaxby's, typical big-box and fast-food establishments in the Deep South, but no country store with knotty pine walls and trays full of flies. Although I felt a bit out of place given the built environment, I knew that, in the nearby southern Blue Ridge, I would find species reminiscent of New England, connecting me, landscape-wise, with my former home in Maine, where I summered in the years before my divorce and where I fished passionately for brookies. In my mind Maine is the real heart of brook trout fishing, and northwestern South Carolina was literally where my southern and northern worlds collided while fishing for brook trout.

In the Deep South, brook trout are relics of the last ice age. Ice sheets that crept into the Mississippi and Ohio River valleys during the Pleistocene epoch pushed these fish, along with spruce trees and sugar maples, south from their northern biogeographical hearth. Plants and animals found today in New England moved into what would become northern Alabama, Georgia, South Carolina, and eastern Tennessee. During the past ten thousand to twelve thousand years, as ice sheets receded and the climate warmed, such individual species as the brook trout, and entire ecosystems associated with colder temperatures, became stranded at higher elevations or disappeared. The moment when I catch a wild brook trout in northern Georgia or South Carolina is splendid, because I am holding a species whose presence is linked to a time when mastodons roamed North America. I feel a sense of amazement when I consider all the geological events that had to occur in order to place that colorful fish here so far from its modern cold-climate environment.

The Chattooga River's watershed is the center of brook trout territory in South Carolina. The river forms part of the border between South Carolina and Georgia and was the fictional Cahulawassee River in James Dickey's novel *Deliverance (1970)*. The area today is dotted with state parks, national forests, and small towns, giving the visitor little insight into the inspiration for the novel and its

disturbing events. In this part of the state, the river is cold and clear, with deep pools, white foamy rapids, and small cobblestones, not a clay-stained meandering waterway such as those on the nearby coastal plain. The river drains a 180,000-acre basin, including parts of South Carolina, North Carolina, and Georgia. Of that vast acreage, 122,000 acres are in three national forests—the Nantahala in North Carolina, the Chattahoochee-Oconee in Georgia, and the Sumter in South Carolina—so much of the Chattooga's watershed is protected.

Before the nineteenth century, the Chattooga area was part of the Cherokee Nation, which at its height encompassed 135,000 square miles of territory in southern Appalachia. Accounts by European travelers, including William Bartram during the early nineteenth century, noted the presence of large and permanent Indian settlements. The Cherokee practiced intensive agriculture in the rich alluvial soils found in river floodplains. So people have altered the region's forests for many centuries and continue to alter them more indirectly through invasive insects, including the wooly adelgid, and climate change. Although no documentation exists, natives no doubt caught, netted, and speared brook trout, creating their own relationship with local streams as a home water.

The effects of native farming were lost on me as I hiked and fished in the national forests. As I looked around from the banks of the Chattooga on my first trip, the landscape seemed stable and permanent. That perception was, of course, naive, which I realized as soon as I thought a bit more closely about what I had seen. With or without people, nothing in nature is ever static: trees fall, fires affect watersheds, gravel bars move, and boulders tumble downstream during spring floods. But the regenerated forests concealed their history well. During my spring visit, migrating warblers passed through the canopy above where I fished. I heard several different calls but was able to catch only a brief glimpse of the Blackburnian warbler and its glowing plumage as it flew to breeding grounds in the mountains that house high-elevation boreal forest in the eastern United States and eastern and central Canada. I was following that bird north as I fished my way to Maine. I was happy to see that specific species, because the bright orange throat of the male reminds me of the similarly brilliant orange belly of a spawning brook trout.

Despite the native farmers' changes, the area's forests have retained great biodiversity, given the mountainous topography and prevailing weather patterns, including moisture-rich air masses that stream up from the Gulf of Mexico during the hot summer months, delivering their punch-in-the-face humidity. The area annually receives more than eighty inches of rain and snow in normal years. This life-giving precipitation feeds brook trout streams through surface runoff and, most critically, groundwater throughout the hot summer months. Today, even though the area remains biologically diverse overall, South Carolina's wild brook trout can be found in only sixteen streams, or about fifty-nine miles, making up a tiny piece of South Carolina's geography, so my initial skepticism of trout in South Carolina wasn't totally unfounded.

Because South Carolina is located at the southeastern end of the brook trout's range, many streams are fairly marginal habitat even without the threat that climate change poses. But warming waters and restricted habitat haven't limited efforts to protect them. South Carolina, along with neighboring Georgia, has proactive private, state, and federal programs to protect and even expand brook trout waters. The enthusiasm for trout was evident during a chapter meeting of Trout Unlimited (TU) I attended during a second visit to Clemson University, about an hour's drive from the closest brookie stream. The room was packed with members, most of whom had donated time or money to the local brook trout cause. In conversations that evening, members told me their desire to restore brook trout was not necessarily an effort to create a new recreational fishery but, instead, to restore a native species to its rightful place in highland streams. I certainly wasn't the only person in the crowd that night who was passionate about brook trout.

Drought and record-setting temperatures during summer in the Southeast, and in South Carolina in particular, can present major obstacles for those working to protect and restore brook trout. The visible result of the 2006–9 drought that enveloped the Deep South was dramatic as I passed by reservoirs in Alabama, Georgia, and South Carolina. The lakes I passed had a climate-driven apocalyptic look to them. Docks on the region's reservoirs were languishing dozens of yards from the dropping water line, exposing the skeletal shoreline. Expensive ski boats sat idle in yards or parking areas waiting for a deluge, and property owners along the shores of these reservoirs had, ironically, gained quite a bit of real estate, at least temporarily.

While brook trout streams are found far upslope from these diminished reservoirs, the barren shorelines tell a story about similar conditions higher up in the watersheds that feed these reservoirs. In recent years, these tepid lakes have held less, partly because less rain has fallen in higher elevations. Wild brook trout habitat is found exclusive in small streams at elevations as low as 1,700 feet but in South Carolina generally above 2,000 feet. Streams on the low end of this elevation can be considered the most vulnerable, given the increasingly flashy rainfall patterns and record-high summer temperatures. Brook trout deal with drought cycles by holing up in deep pools fed by cold groundwater and springs. So several years of low rainfall does not necessarily spell doom for these ice age relics. That's what I told myself as I tried to remain optimistic while I climbed and sweated in my effort to reach a brook trout stream in the hot early fall during a return visit. What I do fear, though, are longer-term climate shifts, such as hotter, longer, flashier, and ultimately drier summers that might further constrict the range of brook trout in South Carolina.

When I first visited South Carolina's brook trout streams, the water was obviously low, even to someone who had never seen these streams before. Previous high-water marks provided glimpses of better times for trout. As I walked toward a stream, the forest floor felt and sounded brittle. My local TU contact and informal tour guide, Dave Van Lear, a retired professor of forestry at Clemson, seems as comfortable casting on a stream as he does teaching about brookies and serves as the South Carolina brook trout coordinator. He said that the streams we visited were lower than he could ever recall. Van Lear and others involved in the restoration projects work with a sense of urgency, because they have fished the area long enough to recognize stressful conditions. While I always carved out time to fish in each of the states I visited, I also met with individuals such as Van Lear who were directing brook trout restoration efforts or guides who were especially knowledgeable about brook trout. They often were associated with the Eastern Brook Trout Joint Venture and were helpful as I sought to better understand the local plight of brook trout. One important thing I noticed during my travels was that the streams that volunteers help restore become their de facto home waters, even if they don't actually fish them much. They achieve home water status because the volunteers spend so much time and energy making them ecologically whole again with the restoration of native trout.

The future of brook trout populations in South Carolina, as in so many of the states I visited during my trout journey, holds reasons for both optimism and pessimism. Almost all the water at elevations high enough for brook trout is in areas that provide some form of protection, including Sumter National Forest, Jocassee Gorges Management Area, and Table Rock State Park, all found in the northwestern quarter of the state. So these areas will see no wholesale clear-cutting, mining, or other destructive developments in the future. Unlike in other states such as Maine, where private landowners hold so many cold-water resources, fisheries and conservation officials in South Carolina are able to work in a landscape that is already protected and one they have some control over. This allows cold-water conservationists to propose and carry out more aggressive habitat improvement programs, such as bank stabilization or installing fish-friendly culverts, because they are working directly with state and federal biologists (often one and the same) and not private landowners.

While much of highland South Carolina is now protected, historically it was not valued or managed as a conservation landscape for trout. Federal designation of the area as national forests didn't occur until the mid- to late 1930s. By then these lands had been logged over, burned, farmed, mined, grazed, eroded, and sometimes abandoned after the soil washed away. If you look closely, the scars of those land uses are still visible. Along a stream in the now-mature Sumter National Forest (designated in 1936), I have spotted trenches or large gullies dug into the forest floor that served as rough chutes for hauling logs out of the forest during the 1920s. In the same area, I hiked along an old logging road partially choked with rhododendron; ironically, that road provides access for today's hikers, birders, and anglers.

As I stood in that forest, I found it was easy to see it as undisturbed, as old woods, because that's what the area is becoming. The trees were tall, the streams were shaded, and brook trout swam in small pools. White pines and decaying hemlocks reached above the oak canopy. This is not, however, a pristine landscape. When one examines the forest more closely, signs of the hands of humans are never far. Even so, it is not possible to imagine how the forest might have looked a hundred years ago. Old photos show a once-broken landscape devoid of trees and full of often-marginal farms. We live in the moment, so understanding long-term environmental change while standing on a stream bank for an afternoon is difficult or maybe impossible. The forests in most of the eastern national forests today are testament to the resilience of nature if sometimes just left alone to heal.

Today brook trout represent only about 11 percent of South Carolina's wild trout resources. So even within all cold-water stream environments where brookies could exist, other trout species have displaced brookies in the overwhelming majority of these streams. I realized that, as a result, wild brook trout would probably never become a focal game fish species simply because they don't occupy enough geography. Even so, those South Carolina streams resemble other brook trout landscapes with their dark rhododendron and hemlock forests. The landscape also smelled the same. The earthy rotting vegetation and sweet rich scents of pine and hemlock made it easier to shut out the reality of the brook trout's precarious situation in South Carolina.

As in so many other states in the eastern United States, the decline of brook trout in South Carolina is largely the result of the introduction and success of brown and rainbow trout and of changes in habitat that lead to warmer waters. These introduced species have displaced brook trout up to waterfalls and other physical barriers that limit upstream movement. In South Carolina, brook trout are almost exclusively found in tiny headwater streams. While removing browns and rainbows from larger streams with existing excellent habitat would be ideal—larger streams usually contain more structural complexity as well as more potential food sources—preventing adaptable competitors from moving into these streams from the Chattooga River system below the waterfalls would be impossible. For example, on one target stream, stocked browns and rainbows were removed above a seventy-foot waterfall. Below the impressive falls is no physical barrier to prevent browns or rainbows from recolonizing from the Chattooga. So the brookies get the highest part of the high stream. But this high habitat supports fewer insects and fewer stream structures such as downed timber. Less habitat and structural complexity means fewer and smaller brook trout.

While Van Lear was showing me around South Carolina's brook trout landscapes during my first visit, I was amazed how far upstream brown trout had traveled before a brook trout restoration project removed them. The professorial Van Lear told me that the project had removed a fourteen-inch brown trout from a stream that was literally small enough to jump over with ease. I found it almost unbelievable that such a fish could survive on the meager resources in that environment. But as Van Lear pointed out, that big brown was probably feeding on remaining brook trout. This was the first of many examples of the adaptability and competitiveness of both brown and rainbow trout at the expense of brook trout that I saw during my travels.

Restoration efforts in South Carolina have consisted mainly of removing nonnative trout from headwater streams and improving trout habitat by dropping timber across streams to create deeper, protected pools. While these two interventions might seem simple, they are critical if brook trout are to hold on at the southern end of their range. The benefits of removing nonnative competitors and predatory species are obvious, but the importance of cutting and dropping trees across streams is an often-overlooked component of brook trout conservation. Pools created by the "chop-and-drop" program trap organic material such as leaf litter that in turn creates habitat and provides building block nutrients for microorganisms, insects, and small fish. In other words, these trees jump-start the food chain in normally nutrient-poor highland streams. Downed timber also helps to create deep pools that act as sanctuaries during hot and dry months and seasons, and they collect gravel and other coarse sediments that provide spawning habitat for brook trout. This was not a mechanism to improve fishing per se, at least in the short term, because an angler has no easy way to cast in that riotous tangle of brittle branches. Instead, as I stood on the scaly trunks of one of the downed trees above the stream, I had to simply admire the newly created habitat.

The chop-and-drop program has become the cornerstone activity for improving brook trout habitat in South Carolina, but it is not as simple as it might sound. If trees land in the wrong spot, they won't create pools; instead, they will expose the stream to more direct sunlight and thus higher temperatures. So an experienced timber cutter is a prerequisite, and the work is dirty, hard, and precise. Without labor provided by local Trout Unlimited volunteers, brookies would today be in a much more precarious state in South Carolina. Before that first trip, I had not appreciated the role of TU volunteers in grassroots conservation and habitat restoration. In Tuscaloosa, I live more than five hours from the closest wild brook trout stream, so my perception was based on mere glimpses, a day here and a weekend there. I never had a chance to witness the heavy lifting that produces trout in many locations.

In addition to highlighting efforts to protect brook trout in the eastern United States, one of my goals throughout my travels was to catch a native wild brook trout in each eastern U.S. state that is participating in the Eastern Brook Trout Joint Venture. Given the limited geography of brook trout in South Carolina, I felt this state would pose the greatest challenge. Most states have hundreds, if not thousands, of miles of brook trout streams. It wasn't hard to check off Virginia and Maine, where brookie water is seemingly endless, but given my limited time and South Carolina's limited streams, the state did not provide many opportunities.

Brook trout streams in South Carolina are fairly accessible from U.S. Forest Service roads, but fishing them is another matter. Like so many headwater streams in southern Appalachia, many of South Carolina's streams are literally encased in thickets of rhododendron. Van Lear concurred. "These streams will probably never attract a lot of fishermen," he observed during my first visit. "They are just too damn tough to fish." While that is good for fish, it is bad for someone with a brook trout obsession. Old rhododendron thickets are tantamount to a vegetative obstacle course. Branches grow in all directions, making it nearly impossible to walk through them, especially with a long fly rod in hand. One literally has to crawl through, over, and under the web of branches—an exhausting, dirty, and bewildering experience. It was almost comical at times. Van Lear and I once entered the shrubby maze of rhododendron in Sumter National Forest and had a hard time finding a way out. Blue jays and crows mocked us from the forest's canopy. We laughed because we were lost only a few hundred yards from the parking area. A path will suddenly disappear into a tangled mess of branches, with no apparent way through or around it. Casting to fish is impossible in this jungle, because there are few spots to make a back cast of any distance without the fly's disappearing among the branches. I continue to be surprised that I have never broken any rods while crawling or casting. But, to be safe, I usually fish these areas with my most inexpensive rod.

Fishing in streams mummified by rhododendrons isn't really fly-fishing. A better term is fly acrobatics. Given that flies have so little weight, it is difficult to place them in tight spaces. During that first trip to fish a stream in South Carolina, I literally had to drape a weighted wooly bugger over a branch in an attempt to catch a small trout hovering under a tangled mess. After a brief inquisitive approach, the brookie turned away, never to be seen by me again. As I raised my line, the fly wrapped around a small branch, leaving me to wonder whether I really needed to catch a brook trout in every eastern state. I had spent an hour crawling through the mess to reach the stream, and in an instant my fly was gone, the fish was gone, and I was suspended in a web of branches, contemplating my next move. Whoever said brook trout are easy to catch has not fished in a southern Appalachian maze of rhododendron. Given these challenges, my first attempt to catch a brookie in South Carolina was an abject failure. Brookie fishing in such an environment isn't contemplative. No long elegant casts. No Zen rhythm. The shrubs see to that.

My shining moment came on my third visit to South Carolina. Yes, it took three visits to catch a brookie! But these trips weren't much of a sacrifice. The clean cold rivers and mature forests make the area a pleasant part of the South. And while I enjoyed spending time with local guides and TU members, I also wanted to fish alone. I wanted to experience the landscape and stream on a more personal level, without the distractions of other people, even those I considered friends. It is one thing to be shown and told about a landscape, but being in the stream, listening to and watching the landscape, is also imperative.

I hiked up Crane Creek in Sumter National Forest, fishing small pools created or deepened by downed trees. At the first sizable pool, I was able to find a bit of a clearing to cast from far enough away to avoid spooking the small fish. A handful of small caddis flies flitted above the stream, and I hoped they would draw the attention of the fish below. I flipped my elk hair caddis to the base of a small waterfall and watched it slowly drift back and around to slack water. As soon as the fly reached the quiet water, wham! A tiny brook trout lunged out of the water as it ate my fly.

A brook trout strike is always impressive for its sheer enthusiasm, no matter the size of the fish. I had never played a small fish as carefully as that one, because I needed to place South Carolina on the "caught fish" list. Because both the fish and the pool were small, the fight lasted all of about ten seconds, but it was memorable. After a quick photo to commemorate my prize, I held it gently in the palm of my hand and watched it dart back into the deep end of the pool. On my first cast of the day, I had landed a South Carolina brookie. It seemed so simple after the fact.

By that time in my life, I had caught hundreds of brook trout, so it wasn't a challenging angling skill per se. Instead, what I felt was relief that I had caught a South Carolina brookie among the rhododendrons. I felt a little more grounded in the Deep South for having succeeded at my brook trout quest in South Carolina. After bushwhacking downstream, I finished the rest of day on the wide-open Chattooga, where I could exercise my casting arm. But even while making long fluid casts on the open river and catching several stocked rainbows, I felt more satisfaction with the small brook trout I had caught earlier in the day.

On a stream as small as Crane Creek, not more than ten feet across at its widest, fly-fishers have no expectations

Caddis fly (*Nyctiophylax*).
Drawing by Frederick Steinberg.

of catching large trout. On a certain level, therein lies the joy of brookie fishing. With no trophies to catch, fishing greed doesn't take hold. The shrubs and stream keep your fishing ego in check.

Fish size, of course, depends on geography and habitat. Brook trout in the South Carolina and Georgia mountains of southern Appalachia are generally small, because they are limited to headwater streams with less food and cover. Farther north, brookies live in larger streams, rivers, and even lakes, all offering a wider assortment of such high-reward food items as forage fish. Scale of habitat thus corresponds to the size of the fish. It is a revelation, though, to know that the first six-inch brook trout I caught in a foot-deep clear stream in South Carolina is the same species as the giants I would later catch in Labrador.

In Buddhism, the Second Noble Truth states that the origin of suffering is desire. I often think about this while angling, because anglers typically want more and bigger fish and are disappointed if they don't catch those fish. One day, a twelve-inch trout is cause for celebration, but the celebrations quickly diminish if bigger and better fish don't follow. My desire is usually kept in check on small brookie streams, because, even if I fish perfectly, most streams cannot grow fish larger than eight inches. So instead of being consumed with an imagined trophy, I can hike, watch the stream, listen to the forest and birds, catch fish, and, on a good day, enjoy the surroundings, although the rhododendron groves make it hard to enjoy or see anything at times. Make no mistake: at times (some of which I discuss in this book) I became greedy for bigger fish. But for me, brook trout fishing provides a complete experience in nature in contrast to fixating on trophy fish. There is rarely a day on a brookie stream when I lament the one that got away.

CHAPTER 3

Tennessee

ALONG INTERSTATE 59, billboards for Rock City and Ruby Falls, longtime tourist destinations, were telltale signs I was getting close to the Smoky Mountains and brookie country. As I followed the interstate north out of Tuscaloosa, where the southern Appalachian hills begin, and drove toward Chattanooga from Birmingham, over the fall line and into the southern tip of the Appalachian Plateau and its distinctive ridge lines, I-59 slowly but steadily climbed higher, the hills around me got bigger and steeper, and they grew more oaks, hickories, and Virginia pines. In addition to the trout that awaited me in the mountains of eastern Tennessee, my drive toward Chattanooga on that particular trip in 2015 signaled the beginning of my annual trip to the Maine coast and my family during summer break. (Such trips have ended since my divorce.)

Even as a young boy traveling from our home in St. Louis to Florida's Gulf coast on annual family vacations, I found those distinctive billboards—the classic white-on-black VISIT ROCK CITY sign and the red-on-white SEE RUBY FALLS sign—excited me because they represented the entryway to a new landscape. I felt we were finally making

progress on our way to the beach by way of Tennessee. Gone were the monotonous cornfields and rolling farm country of Missouri. Now we were approaching the mountains, and after the mountains came Florida and fishing its Gulf waters. Today that same childhood sense of anticipation of new country enters my mind when I head north on I-59 and see the first signs for Ruby Falls and Rock City. Gone are the monotonous pine plantations; now I am in the foothills of the southern Appalachians, and after those hills are the Smoky Mountains, and in those mountains are trout.

Given that it is only a five-hour drive from my Tuscaloosa home to the mountains where brookies live, I have spent more time on brookie streams in Tennessee than in any other southern state. Because I live in bass country, Tennessee represents a brook trout lifeline. I am fortunate that Tennessee contains two large brook trout landscapes: Great Smoky Mountains National Park and Cherokee National Forest, which extends into western North Carolina. Some streams in Great Smoky Mountains National Park have sizable fish, and they even allow room for casting, which makes fly angling even more enjoyable. The park has plenty of rhododendron-entombed small streams, but if you are willing to hike, climb, bruise, and even bloody your knees getting over boulders, isolated streams with deep pools wait to be fished, many of them full of brook trout. And given that these streams are in both a national park and national forest, many are largely protected from the direct impact of people, in contrast to brook trout streams outside these areas that face monumental challenges.

When I visit Great Smoky Mountains National Park, I usually stay in Townsend, near the southwestern side of the park, about an hour from Gatlinburg. Townsend is a quiet unpretentious town with just a few hotels and restaurants. Much of the town still has the feel of a 1960s tourist destination. In the off-season—my favorite time to be in the park because of the small crowds—most restaurants have limited hours or are closed, making Townsend feel somewhat deserted. But even in this quiet corner of the park, change is coming. More vacation homes, hanging precariously to the sides of mountains, seem to pop up between my frequent visits. The trend I saw in northern Georgia has spilled into eastern Tennessee.

I acknowledge and struggle with my part in this change, as I do when I visit Georgia. I lodge and shop here, and I use resources, so I realize I am part of the problem. But at least in Townsend, I can stay close to the park, buy groceries and cook for myself, and try not to demand too many new services. While the landscape has changed, with a few more restaurants and a first-class fly shop (I don't complain about that type of development), the change is occurring on nowhere near the scale of what is happening nearby. Just up the road, around Gatlinburg, the problems

faced by brook trout, among countless other species that call the mountains home, become crystal clear.

U.S. 441 between Pigeon Forge and Gatlinburg overwhelms my senses and appetite as I pass through on my way to the park. Cheap corrugated buildings line the highway, which usually has bumper-to-bumper traffic, especially during holidays, the summer, and peak times of fall foliage. None of these structures looks permanent, and none relates architecturally to the natural wonders of the area. This place purports to celebrate southern Appalachia while doing everything possible to obliterate any semblance of natural landscape in sight. Does the planet really need Professor Hacker's Lost Treasure Golf, an Elvis Museum, Dollywood, and the Hoot N' Holler Dinner Show when Great Smoky Mountains National Park, with more than eight hundred miles of trails, is just down the road?

I do not find fault with local people who work in these establishments—we all have to earn a living—but I do blame the individuals who set in motion this type of slipshod development. Most of this growth has occurred since 1982, when Knoxville hosted the World's Fair. Local planners and developers had an opportunity to plan more appropriately, given the unique qualities of the local environment, but they chose the putt-putt golf option.

The tourist developments further isolate the national park and the creatures that call it home. Of course, such isolation isn't limited to this spot; it is happening to ecosystems and individual species throughout the crowded East and especially in the southern Blue Ridge. But this isolation is unique in that it is happening to a national park in one of the most biodiverse regions within the United States, rivaling the Tensaw River's ecosystem near Mobile, where the famed biologist E. O. Wilson grew up and learned about nature. When populations become isolated, when their range is reduced, it is usually a matter of time before that population disappears. And so the park is losing its environmental connections with surrounding areas, affecting not just trout but salamanders, insects, birds, and bears. Genetic isolation is a dead-end street for most species.

The park's enormous acreage in the southern Appalachian Mountains—the Blue Ridge, to be more precise—is unparalleled in terms of diversity in North America: more than 17,000 species have been documented here, and many more likely remain undiscovered; about 100 species of native trees, more than in any other national park; more than 1,500 flowering plant species; more than 200 bird species; 66 types of mammals; 50 native fishes; 39 varieties of reptiles; and 43 species of amphibians. While much of the park was logged or farmed or both, the National Park Service has estimated that about 150,000 acres have never been cut or cleared. No other area on the planet of equal size in a temperate climate can match the amazing diversity of plants, animals, and invertebrates.

The roughly southwest-to-northeast geography of the park is the driving evolutionary force behind this diversity. Elevations within the park range from 876 to 6,643 feet, with unbroken mountains that rise more than 5,000 feet for more than thirty-six miles. This range in altitude mimics the latitude. Thus, as I climb higher on a stream, I am hiking toward Maine in terms of flora and fauna. At the top, I am in alpine red spruce and Fraser fir forests. Ecologically, I am in New England and southern Canada.

Because I am in search of brook trout, I am for the moment more concerned about the impact of local develop-

ment on that species. What few brookie streams exist outside the park are surely on their way out, a result of the massive landscape changes taking place. Development in the East is unavoidable in such desirable landscapes as the southern Smoky Mountains, but we can, or at least we should, have smart development. Because of its proximity to the park, Gatlinburg at least offers outdoor activities, including fly-fishing, river rafting, nature tours, and horseback riding. But even here, the tentacles of strange tourism unrelated to the park reach deep, with a Ripley's Believe It or Not! Museum located just outside the park's boundary. I have grown to love the park so much that I find it difficult to say "to each his own" regarding entertainment choices.

What does this concentration of people and their environmental impact do to the park and its surrounding areas? Polluted water, exhaust from vehicles, trash, and runoff from parking lots and roads all have a detrimental effect on the local environment. During one visit, I pulled off the road and scrambled down the bank of the Little Pigeon River to check the condition of the water. This river follows the congested road between Gatlinburg and Pigeon Forge, so I suspected it would be heavily polluted. Plastic grocery bags hanging from skeletal winter shrubs fluttered in the wind like sinister prayer flags. Beer bottles, hubcaps, and Styrofoam cups also littered the banks. The water was clear, because it was flowing out of the park, but the litter ruined the stream overall. The local Trout Unlimited chapter has annual river cleanups, but it's next to impossible to keep this area clean, given the traffic flow. A friendly passerby stopped and asked if I needed any help. When I told him I was just looking at the stream, he laughed and quipped, "The trout are in the park, son."

Once I shook off the traffic jams and residue of Dolly-wood, I returned to the park and its trout. The park's populations of brook trout are some of the most studied in the eastern United States. The research by park biologists has been ongoing since the early 1970s. Specific research questions change through time, but continuing efforts, according to Matt Kulp, one of the park biologists, include monitoring both water quality and brookie populations. I met with him at the park's headquarters near Gatlinburg in 2009. With stonework and a heavy wooden front door no doubt made of oak, the building is a classic 1930s Civilian Conservation Corps structure. Inside, the building featured coffee-colored wood paneling and flooring. It just felt like a National Park Service building. I expected nothing less. His basement office was a cramped space for someone who looks like he could have played linebacker in college. There were stacks of paperwork on his desk and field equipment piled in the corner, confirming that he has a lot of projects to oversee. He struck me as someone who is more comfortable in a stream than in a basement office.

Specific research taking place in the park includes mapping the genetic variation of brook trout populations to determine how diversity varies from stream to stream. This is important, because, if great variation exists, it will affect how restored streams are stocked. To the extent possible, biologists want to prevent stocking mixed populations, if they show significant genetic variation; the biologists want to maintain the evolutionary history of stream A by stocking it with only stream A fish. Park biologists are also monitoring water quality, and they are especially concerned about impaired or polluted streams, which, in this case, refers not to trash or sewage pollution but to pollution caused by acid rain and acidic sulfates from coal-fired power plants that have built up in the soils and continue to leach into streams. At least Twelve streams in the park have water quality poor enough to warrant designation by the Environmental Protection Agency) as impaired in 2022. Even if all acid rain stopped today, streams would continue to be impaired because of the large quantities of sulfates stored in the soil.

I asked Kulp if the park service has any liming programs intended to mitigate the effects of acid rain in streams, and he told me that such programs are impossible, because human intervention in a national park is highly restricted. I found this to be a catch-22, given that the streams in the park are protected, but if a threat develops from outside the park, federal law limits what the park service can do to mitigate the threat. Kulp agreed with my assessment, but the rules for altering in-park environments are strict, even if intervention might improve a specific situation. He went on to say, "There are good reasons for such a rule, but it can handcuff recovery efforts in certain situations."

Acid rain is deadly to fish, because, as it travels through soil, it frees up aluminum and other metals that are naturally in the ground. As this free aluminum enters headwater streams, it poisons the streams' creatures. When a trout's gills come into contact with free aluminum, the fish's blood vessels burst, literally causing a sort of fish heart attack. From its banks, an acidic stream may look healthy—the waterfalls are pretty, the deep pools look like perfect trout habitat, but the water is toxic.

In addition to running their monitoring programs, the biologists have removed nonnative trout from specific streams to create more favorable conditions for brook trout. As has happened elsewhere, these interventions have met with public concern and, in some cases, anger, but Matt noted that the park has a mandate to not only protect native species but to remove nonnative species whenever possible. It is a national park after all, the paramount American conservation institution and geographical unit. The public concern was not about the use of piscicides in principle; rather, people were unhappy about the removal of all the browns and rainbows from target streams. Piscicides inhibit a biochemical process at the cellular level that prevents fish from absorbing the oxygen in their blood, thereby suffocating them. From a conservation angle, Kulp predicted that Lynn Camp Prong, one of the treated streams, would eventually develop into a blue-ribbon brookie stream, given its size and in-stream structures. This stream reopened in spring 2015, giving brookies time to become established in their new rainbow-free home. And Kulp was right: today the stream is one of the best for trout fishing in the area. Overall, about seventy miles of streams in the Tennessee section of the park are considered brookie streams, certainly not an insignificant number, but this pales in comparison with what existed 150 years ago.

Brook trout have been studied intensively and so much restoration work has been directed their way in the park for good reason: since 1900, its brook trout have disappeared from 75 percent of their original range. So the research is a true ecological necessity. As in other states I visited, brookies have been driven out of the larger, lower-elevation streams and into the high marginal headwater streams that offer little space, cover, or food. Early on, logging, farming, and the stream destruction that followed forced brookies higher in the watersheds; after forests regenerated, the presence of the more competitive nonnative rainbow and, later, brown trout were the main culprits in their decline in more recent years.

As I fish and hike in the park, I often find it hard to imagine how different the present-day landscape is from what it once was. Today, where hundred-year-old beech, hemlock, and white pine tower over trails and streams, historical photographs show clear-cuts, ragged cabins, and families eking out a living on small farms with more rocks than dirt. If you look closely today, you still can find the imprints of those former residents.

While I was fishing up LeConte Creek one November in a rehabilitated brookie stream that had all its rainbows removed a couple years before my visit, I suddenly came upon a rough stone wall in the forest. My son Frederick, on a visit from Maine, was completely unaware that the area was ever anything but a forest and wondered how such a wall was built and by whom. I told him that the land was not a park a century ago but was filled with small farms and large logging operations. I was not sure he believed me until, later that day in the same area, we found a cabin, preserved by the park as a historical museum, that probably was home to the folks who'd built that stone wall.

The idea of home water in the park has changed during the past century. Before Congress created the park in 1934 and its official dedication by President Franklin D. Roosevelt in 1940, local folk fished and depended on these streams for part of their sustenance. Although few early accounts of fishing survive, the ones that exist describe bountiful fishing for both native brook, or speckled trout, as they were called locally, and the introduction of rainbow trout. In *Twenty Years of Hunting and Fishing in the Great*

Smoky Mountains (1935), Samuel J. Hunnicutt describes regularly catching more than fifty trout during a single outing. But even by the 1920s, Hunnicutt notes, he had to hike to high-elevation streams to catch brook trout, because they had disappeared in lower-elevation streams.

After the national park was created, the concept of home water began to change from that of locals who fished for food (and previously lived in what became the park) to visitors who fished more for recreation. This shift was initially slow, as locals continued to exploit the park's fish and wildlife and ignore the new regulations, but the shift to recreation began nonetheless. The same transition took place in many national parks and national forests throughout the eastern United States, as mountain folk were forced out and tourists came in.

According to early travelogues about the area, such as John Preston Arthur's *Western North Carolina History from 1790 to 1913* (1913), rainbow trout were already well established in the park by the eve of World War I. Then as now, finding rainbow trout in the park's streams was not difficult. In fact, if you fish just about any stream below thirty-five hundred feet, you probably will catch this incredibly adaptable fish, whose native waters are on the west coast of North America. I find it frustrating to hike and climb up, pool after pool and boulder after boulder, and not catch a native brookie.

Today the park's only rainbow-free streams are those that were too remote to reach with the stocking truck years ago or those sections of streams above waterfalls too high for introduced trout to reach. Park officials stocked millions of rainbows during the twentieth century, a management practice that finally ended in 1975. By then, though, the seeds of the rainbow's success had been sown, because it would prove impossible to remove them from all streams in the park. For most of the last century, native fish species took a backseat to recreational fish species.

In my quest to catch brookies in the park, I sought the help of Ian Rutter, a fly-fishing guide based in Townsend whom I tracked down through his informative website; it provides up-to-date information about fishing conditions as well as fantastic photographs of the park's trout. The photos are what lured me in. I had caught plenty of tiny brookies on miniature streams in the past, but I wanted to fish with a local who could take me to a bigger stream with more robust fish. I also like to fish with guides to hear their thoughts about the state of brook trout in their home waters.

The morning we set out, the wiry Rutter hiked like a man on a mission, eager to leave the parking lot and picnickers behind, while I struggled to catch my breath in what felt like the high mountains. Of course, he did this for a living and was younger than I, so I did not feel so bad. I think he was trying to gauge my fitness level, because the stream we ended up fishing required both hiking and bouldering skills. As we scrambled down a sheer entryway to the stream, he said, "Brookie fishermen are usually different, more interested in the overall experience of being on a hard-to-reach remote stream. So size of the fish isn't always the most important thing." I felt like I had found a kindred soul.

In Rutter's home water of the park, he knew exactly where brookies began to dominate. The unnamed stream we fished after a few miles of hiking upslope had plenty of deep black pools, cascading falls, smooth gray boulders with moss beards seemingly tossed about, and towering birch, beech, basswood, and hemlock, all set in a steep canyon. It seemed to me there were only a few ways in and out of the gorge, so I was lucky to have a local show me the way. This was not the typical southern Appalachian brookie creek with a cast-preventing rhododendron veil. Instead, the stream had plenty of room and many perfect pools.

Conditions were just fine: no other people and a bounty of caddis flies luring trout from the depths of their pools. My fishing luck was magical. Some days, I miss strike after strike while fishing dry flies. In frustration, I have blamed a lack of coffee or too much coffee, aging, sinning, and karma. Anything but me. Guides roll their eyes when they hear their clients' excuses for failure or, worse yet, when guides get blamed for a client's poor mechanics. But, on that day and on that stream, I had a 90 percent success rate: for every ten strikes, I caught nine fish. My guide commented, "You're on a roll."

My success was partly thanks to Rutter's exceptional ability to spot fish. Every time I caught one, he had already identified the next casting location. Most fish were small, but many brookies were in the eight-to-ten-inch range, big for the Smoky Mountains. I even caught a couple of fat rainbows. It was nice to fish in a stream where rainbows were clearly outnumbered by natives. I cursed the sun's movement across the sky and behind the ridge because I did not want the day to end.

While sitting on the bank eating lunch in the midst of a mayfly hatch, I asked about the state of brook trout in the park. Rutter told me, "They are in good shape, and there is more and more interest in fishing for natives." He also complimented the park biologists' efforts to restore brookie streams, saying, "That is going to provide

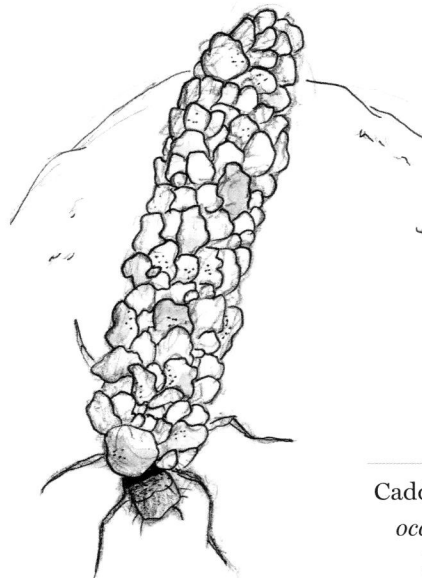

Caddis larvae (*Brachycentrus occidentalis*). Drawing by Frederick Steinberg.

high-quality fishing opportunities in the future. And that is the best way to get people interested in native fish."

The other geographic center of brook trout in Tennessee is in Cherokee National Forest, which borders Great Smoky Mountains National Park on both its northern and southern ends. As in the park, research on and restoration of brook trout has a relatively long history in Cherokee, which was established in 1920. For example, rainbows were removed as early as 1958 because of their noticeable impact on brook trout. That is shockingly early when you realize that, during the 1950s, all other eastern states were stocking as many farmed rainbows as they could hatch, and Great Smoky Mountains National Park did not escape the stocking. The early efforts to remove stocked fish from Cherokee certainly bucked the trend of widespread and

Mayfly (*Heptageniidae*).
Drawing by Frederick Steinberg.

intensive trout stocking so common elsewhere in the eastern United States.

Given that about 70 percent of all brookie streams in Tennessee lie within Cherokee National Forest, perhaps the early attention they received is not surprising. In most states, actually removing trout to enhance brookie populations was not even in the thoughts of fisheries managers. Other early conservation activities in Cherokee included the construction in 1968 of a barrier on Henderson Branch to prevent rainbows from migrating upstream and more rainbow removal efforts during the 1980s and 1990s. Like Great Smoky Mountains National Park, Cherokee is a biologically rich landscape in that it provides habitat for 43 species of mammals, 154 fishes, 55 amphibians, 262 species of birds, and thousands of insects, including a diverse assemblage of aquatic families, among them stone flies, mayflies, and caddis flies.

When I visited Cherokee National Forest on a return trip to Tennessee one fall, I met with Jim Herrig in the southern section near Tellico Plains. Herrig studied under the giant of trout research, Robert Behnke, at Colorado State University, served as a fisheries biologist beginning in 1987 and has written many of the reports on brook trout in Cherokee, so he was another obvious person to track down. Herrig suggested we talk about brookies while fishing, so we set out for Sycamore Creek, one of his favorite brookie streams in the national forest. My visit coincided with the opening of the fall bear and boar season, so as we pulled into the parking lot, we were met by a few grizzled hunters in worn camo coats who had bigger quarry in mind than a mountain brook trout. Nor did they think much of my official University of Alabama car. As we headed toward the streams, I could hear them grumbling about Alabama's Crimson Tide football team. Herrig quipped, "Tough crowd."

Talking while fishing presents an interesting set of challenges, because both parties want to fish, not talk. So questions and comments come between casts and over the roar of rapids. Herrig, tall and thin, is a methodical fly-fisherman; he looks and fishes almost professorially. He slowly moved up the stream, casting to every trout possibility along the way. Given his physical stature and focus, I imagine he could be an excellent tennis player or golfer. I almost always fish alone, so it can be enlightening to watch how someone else, someone with far greater experience than I, approaches a stream. Herrig represented himself well, catching many brookies in small pools and pocket water.

His feelings about the future of brookies in Cherokee National Forest ranged from optimistic, given all the work to improve habitat here, to concern about what warming temperatures will do to streams in the South. Between casts he told me, "We will probably lose some streams this far south, if temperatures increase, especially if you couple those temperatures with less rainfall, like we've seen recently." Later, upstream, he commented, "I suspect brook trout have taken a big hit in a couple streams here that were already really small before this drought." He added, as he held a small brookie he had just caught, "Of course, brook trout have dealt with environmental change in the past, so it might not be all bad news. But if you combine warming with less water, we will have some problems on some of these streams."

Herrig also mentioned the lack of cover, specifically large woody debris, as a potential problem. "If there are enough deep pools," he said, "brook trout can hang on in dry, hot years, but when a stream lacks snags and pools, it will be tough." He noted that the removal of beavers in the national forest is one factor that has resulted in poorer trout habitat. According to Herrig, "When a beaver moves into one of these streams, it doesn't take too long before it gets shot by a local hunter. We need some of those beavers, though, because, if we lose fish [as a result of climate change], it will be on those streams without any downed timber and pools." This example shows the complexity of managing nature. Remove one organism and a whole slew of potential ecological problems might develop, many of which we won't even recognize until the damage is done. "Shoot a beaver, lose a trout" could be a conservation bumper sticker.

Because Cherokee is a national forest, it has challenges related to hunters and such landscape-altering endeavors as timber cutting that state or national parks do not face,

so Herrig and his colleagues at times have to work within interesting social dynamics. But despite the sometimes challenging circumstances, brook trout populations have expanded since the 1980s, thanks to the intervention of the national forest's staff.

National forest biologists in this region also have to deal with the destruction of headwater streams by off-road vehicles. Off-roading is popular in the South, and national forests have allowed and even promoted it on designated trails. Off-road vehicles cause silting in high-elevation headwater streams and destroy bankside vegetation. Trails and nearby streams literally become mudholes. This kills anything that lived in a specific stream, and much of this mud also finds its way downstream, where it silts-in previously clear-running streams with a clean gravel bottom. Brook trout cannot spawn in such silted streams.

The impact of off-road vehicles on trout has been clearly documented in the Tennessee–North Carolina border area. Trout densities in streams affected by off-road vehicles in the Tellico River's headwaters in the farthest reaches of western North Carolina are approximately half of those in similar streams across the Nantahala National Forest in the same region. From 1996 through 2004, annual fish counts documented declining trout numbers because of the off-road trail system, including at least one year when biologists could find no first-year fry at all.

The clash between trout conservationists and off-road enthusiasts in Cherokee and the adjacent Nantahala came to a head in October 2009, when the U.S. Forest Service banned vehicles from trails near the headwaters of the Tellico River, a key brookie spawning ground. The Southern Environmental Law Center called the Tellico area one of the Southeast's largest areas and most intensively used for off-roading, with forty miles of designated trails and an average usage once estimated to be twenty-four hundred vehicles per month.

The decision angered off-road truck enthusiasts. Carla Boucher, legal counsel for the United Four Wheel Drive Associations, stated in 2009, "The Forest Service has failed to make the case with regard to adverse effects and resource damage." But off-roading on public lands, especially ecologically diverse and sensitive public lands, is a destructive and often illegal hobby rather than an inalienable right. My hobby, fly-fishing, is not destructive. It does not, through environmental destruction, eliminate outdoor activities enjoyed by other people. Fly angling doesn't destroy someone's home water. National forests should be available for various types of public use, but it seems logical that these uses should be sustainable. If off-roading negatively affects or even destroys other activities and environments, it fails to be sustainable.

Before I paint too negative a picture of the landscape in Cherokee National Forest as a result of off-roading, I need to note that Tennessee has a dedicated environmental and fisheries community that is fighting to protect the brook trout and educate the public about its plight. In addition to the biologists, many souls—usually volunteers from the five eastern Tennessee Trout Unlimited chapters—work behind the scenes, organizing fly-fishing events such as casting clinics for youngsters, educating school groups about environmental issues, organizing "trout in the classroom" campaigns, and providing the labor crucial for the restoration projects.

Trout in the classroom, for example, is a conservation-oriented environmental education program for elementary, middle, and high school students. The program helps students make the connection between human activities and nature through lesson plans focused on water quality, watersheds, trout habitat, and conservation ethics. This unique program also allows kids to see and study the early life cycle of trout, from eggs to fingerlings, through aquariums in the classroom; students then release the trout into a local stream. This program is building the next generation of cold-water conservationists through experiential learning. Couple this activism with two major brookie landscapes that are largely protected, and the future of brook trout in Tennessee is secure.

As I hiked up Lynn Camp Prong with my son Frederick, I felt good about the future of brookies in the national park and beyond. Here, deep in a valley, I saw no go-cart tracks or jammed parking lots, just forest, chattering warblers, and optimism. It is another world and, for me, a better world than that found outside the park's borders. The stream was still closed to fishing during that visit, but I pointed out a couple of small brookies that were dashing for hiding places as we walked slowly up to a pool. It was great to see these natives in their rightful place.

Frederick barely glanced at the little brookies, as he had seen plenty of brook trout while tagging along on this and other trips. We sat on the bank in silence for several minutes and watched as a small trout reappeared and hovered in the current; its cream-colored fin tips fluttered like delicate moth wings and made it visible in the dark water. A better world indeed.

CHAPTER 4

North Carolina

AS I TRAVELED ALONG the Blue Ridge Parkway in western North Carolina, a cool, dense fog hung on the road, and moisture dripped from the overhanging hemlock, maple, and, at the highest elevations, spruce trees. The air was chilly enough that it slapped my face, for I had to drive slowly with my windows down, feeling my way through the misty white muck. No doubt about it, I was in trout country.

Asheville, the regional hub city and my destination one August on my drive back to Tuscaloosa from Maine, draws visitors from throughout the region and beyond with its physical beauty, neat downtown and greenway, access to outdoor activities, and wide-open southern culture. I could live here; it felt familiar and comfortable. Rivers, mountains, rafting, hiking, fly-fishing, microbreweries, a university, great music and art—the Asheville area has it all.

Asheville also has a feel for both mountain and outdoors cultures. Bluegrass music plays in almost every bar in the historic city center, and the town has a couple of fly shops and outfitters. I pulled into one shop near the center of town after the long late-afternoon drive to get some advice and pick up some flies. Fly shops clearly operate on a

quid pro quo system, so I make it a habit never to ask for information if I am not buying something. I do not have a problem with this symbiotic relationship, because fly shop owners won't be around long if all they do is hand out free advice. The presence of a fly shop is a good sign the fly-fishing tribe has a presence in a particular location. In western North Carolina, fly-fishers can drive around with nine-foot fly rods on top of their cars and not really stand out. It's good to be with one's people.

As the sun began to set, I walked to a restaurant for dinner, a few beers, and local music. I found a seat at the end of the bar, my usual spot during solo travels. Bar food— heavy, rich, and salty—washed down with a bitter, hoppy local beer is good stuff at the end of the day. It sure beat the hell out of a granola bar in a chain motel, my usual travel dinner. A local band, Sons of Ralph, worked the crowd into bluegrass frenzy. Young and old, men and women, urban and rural, washed and unwashed spun, skipped, and clogged on the dance floor. The sweet smell of patchouli traveled from the dance floor like a thick blanket of mountain fog.

It was clear why Asheville has developed its reputation as a hippy center of the South: the place was packed with pretty young women in flowing flowery dresses. I finished a last pint, bought a CD from the band, and headed into the warm muggy night. Fly-fishing and hiking up steep streams requires a clear head, and I already knew I was going to pay for the beer, cigarette smoke, and loud music early the next morning. As I walked back to my hotel, dozens of common nighthawks swerved and swooped in the night sky, calling out with their nasally *peent, peent* and catching moths attracted to city lights. I hoped the clouds of insects were a sign of good angling ahead.

The local geography lends itself to trout and outdoor adventures. The mountains mean cold streams and high peaks, including Mount Mitchell in the western part of the state, the highest point east of the Mississippi River at 6,684 feet. The cold fog-shrouded spruce forests around its summit are more reminiscent of Maine than Dixie. I took some time away from fishing during the first part of my several-day stay to hike the Commissary Ridge Trail in Mount Mitchell State Park and could not quite accept that I was not in northern New England. The forest floor was spongy, mossy, and saturated. The spruce trees looked almost black against the battleship-gray sky.

As I moved from brilliant greens and hot summer winds to an almost black-and-white mountain landscape with cold wispy fog, I was shocked out of my late summer mindset and mood. The dull colors, interrupted only by the occasional golden-crowned kinglet flitting about in the trees, gave me a déjà vu moment of being back in New England in early July, when I hiked Mount Katahdin, the northern terminus of the Appalachian Trail. Mount Mitchell felt like a dreamland for August, a welcome respite, especially given the drought and heat that had settled in the South that year. The experience reminded me again that drought is a problem not only in the West.

Reentering the Deep South after spending a summer in Maine was always shocking to my mind and body. Going from sea breezes to the humidity that envelops you like a wet wool blanket is a harsh reentry in August. So the late summer 2016 visit with the cool mountains and spruce trees was most welcome, helping me hang on to Maine a little longer. As I moved south into the heat, I knew my connection with my northern home waters was becoming more tenuous and distant. August was the worst possible time to leave New England.

High up on the ridge line, I did not come across streams of any size, but the cold water seeping through the moss was the source of the many streams farther downslope. This rugged western part of the state is home to hundreds of rain- and snow-fed freestone streams whose cold waters cascade over the rough terrain. Most of these high streams teem with brook, rainbow, and, lower down, brown trout. North Carolina is blessed with extensive high-elevation geography that neither Georgia nor South Carolina has. And the numbers bear this out: state and federal fisheries officials have documented more than five hundred populations of brook trout in thirty-two counties of western North Carolina. So North Carolinian conservationists do not have the same pressure as in South Carolina and Georgia, where the loss of a single brookie stream means the loss of a significant percentage of the state's overall trout population.

Brook trout in North Carolina, however, still face challenges and now live in less than 80 percent of their historic range. This reduction is surprising because their present-day range is still fairly extensive. But you do not have to look far in western North Carolina to understand why brookies have declined. The fly shops, spruce trees, and protected areas here, and the stunning physical geography, attract growing numbers of people and all the services and infrastructure they demand.

The good news is that, today, most known brook trout streams in North Carolina are in protected public areas, including the Pisgah, Nantahala, and Cherokee National Forests; Great Smoky Mountains National Park; and along the Blue Ridge Parkway. The distribution and status of brook trout on private lands in North Carolina is less well known, although state and federal officials continue to survey the state to determine where brookie populations might still exist. It's likely many populations on private lands are not in terribly good shape, given the rapid development of western North Carolina. Many streams around Asheville, for example, once held brook trout, but development, deforestation, and myriad other land-use changes have forced brook trout far into the highlands.

Although brook trout are the only trout species native to North Carolina, wild populations of rainbow trout and brown trout occupy most of the brook trout's former range. Stocked rainbows and browns also have displaced North Carolina brookies, as have northern strains of brookies that were stocked in streams that once contained only southern fish. Studies by state biologists of North Carolina's brook trout populations indicate that 39 percent are the native southern Appalachian strain, 9 percent are northern stock, and 52 percent are hybrids of northern and southern strains.

After a day of hiking and examining the landscape from my car window, I turned my attention to brookies. While it was cool, even cold, on Mount Mitchell, it was blisteringly hot farther downslope on my second day. I was still in brookie country, or at least on a stream where brookies dominated, but I was a few thousand feet below the sweet August air of Mount Mitchell. This was the summer of 2008, near the end of a multiyear drought, and the stress on streams and fish was evident as I fished in the national forest outside Asheville. Even in the early morning, the fish held tight to deep holes and shaded banks.

I could see trout lined up along the bottom, appearing to hover without moving, and, as the water temperature approached 75 degrees on its surface, they were not interested in anything I offered. As I looked at those fish from the bank, I decided not to stress them any more by dragging them up from their cool quarters at the end of my line. So I got back into my baking car and again headed upslope into the national forest to look for cooler water and more eager trout.

What I found instead was a logged landscape, clearly hotter and more exposed without its protective green canopy. Driving into a higher altitude along hemlock-clad roadways and being suddenly thrust into the burning sunshine with downed trees and slash all around me in a national forest was a harsh moment. I felt like I had been thrown into Ernest Hemingway's famous short story, "Big Two-Hearted River." The road I was on had followed a stream, but I suddenly lost it and had to scramble over a lot of dead wood to find it again. My car was covered in dry powdery dust from the parched road. I wanted to see what shape the stream was in so close to the cutover forest. Climbing through a

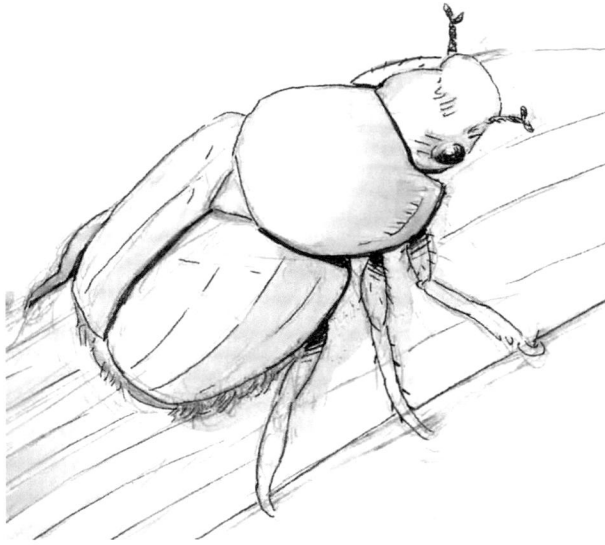

Japanese beetle (*Popillia japonica*).
Drawing by Frederick Steinberg.

slashed-and-burned landscape is a daunting prospect in the South because snakes, some venomous, find such areas quite hospitable. But the sun was hot and bright enough that, like the trout, even the snakes remained hidden in the shadows.

The stream I found after climbing over and through the baked and bleached woodpile was a brookie nightmare: hot, exposed by stumps cut all the way to the banks, and full of chubs. No trout were to be had in this water, whose temperature topped 80 degrees. This was a landscape demonstrated the power of people to dramatically alter nature. No doubt, this forest would regrow and trout would move back in, but the clear-cutting showed the vulnerability of cold-water creatures to poor forestry practices, especially when they are coupled with high summertime temperatures and drought. And the historic increases in global temperature and wild fluctuations in rainfall should mean, in my opinion, some changes to logging practices to further protect increasingly vulnerable cold-water habitats in the South. The question running through my mind that day was whether logging practices would change to protect the vulnerable streams before climate change worsened. Since then, while forestry scientists and the logging industry have debated changes to logging practices and the effects of climate change, the industry and federal government have been slow to change any forestry policy, especially in the eastern United States. The West, with its greater vulnerability to wildfires, has seen some shifts in forestry practices, but these are more related to fire control or lack thereof than to fish conservation.

Because of the heat and drought, I had to travel back up to the Blue Ridge Parkway the next day to find my brook trout in North Carolina. I fished that day with an Asheville-based guide who seemed surprised by my request to fish for brook trout. "You know they don't get very big," he said when we met in an Asheville parking lot that morning. Given the abundance of brook trout waters in the area, I was curious to know if he received many requests to fish for them. He claimed to personally like fishing for brookies, but he rarely got clients who wanted to target them. "No, most tourists want big fish. People don't usually pay hundreds of dollars to catch six-inch fish, native fish or not. You know, dick size and all."

We started at a boulder-laden freestone stream right off the parkway. It was still early enough that the fish were willing, and a bright green floating Japanese beetle did the trick. Beetle, in this case, was a speck of green foam rubber tied around an even smaller hook and did not resemble any insect I had ever seen.

This was typical brookie fishing: I was moving from pool to pool with aggressive strikes in just about every hole bigger than a kitchen sink. The fish topped out at eight inches, impressive given the small pools in which many were hiding. As the day wore on, dozens of people invaded the stream the guide and I were fishing. For me, fly-fishing is almost always a solitary endeavor, and I do not like to see other anglers, let alone throngs of tubers. A Zen rhythm or focus is impossible amid a crowd. By midafternoon, all the plunge pools had become swimming holes, and my angling day was over.

I had caught and released my share of brookies, and, given the heat, I did not blame folks for wanting to cool off in the stream. I was fishing near a roadside parking lot on a hot but beautiful summer day. I had known my angling time was limited, but before quitting for the day, I had to show off a bit and catch a fish I saw in a deep pool next to a picnicking family. After enticing a nice brookie to eat my fly in front of the family, my ego was a bit deflated when the mother declared, "Oh, look at that tiny fish!" She had no idea that an eight-inch brookie is a big fish in these mountains. I did not bother to point that out and let the family return to their picnic. Given the rising temperatures and that the stream was now a swimming hole, the guide and I called it a day and hiked back to the parking area. Fathers sipping cold sweating cans of Budweiser amid coolers and lawn chairs were keen to know about the state of the local fishing. They patiently listened to my descriptions of the small brookies, while their children beat each other with brightly colored noodles.

While my guide claimed he enjoyed brookie fishing, he thought it absurd that larger rainbows and browns should be removed to make room for brookies. In his mind, the stream was no better with native than nonnative fish. "I can't believe some of the Trout Unlimited people want to take trout out of these rivers," he told me over a cold beer on the tailgate of his truck after fishing. He then asked, "What the hell is wrong with the rainbows we got around here?" This opinion is somewhat understandable from someone who makes his living helping people—usually novices—catch fish. While native fish might be important to biologists concerned about the overall ecological integrity of a stream, guides live with more pressing economic realities. I learned that home water doesn't always mean native fish. Few, if any, guides get rich from their trade. The guiding life is a common fantasy among fly anglers, but it is a challenging way to make a living with little room for error. That is not to say no guides are concerned about native fish; guides support many restoration efforts. I understood this guide's point, but I also believe there has to be room for native fish.

While brook trout in North Carolina remain common in various protected areas, why their range has dramatically decreased during the past century, especially on private land, is not hard to see around Asheville. Certainly, western North Carolina suffers from acid rain; the forests high in the Smoky Mountains and around Mount Mitchell bear the scars of air pollution. Introduced species such as rainbow and brown trout compete for, and eventually dominate, the best cover and feeding pools. And rising temperatures pose a critical threat to brookies, given that North Carolina lies near the southern end of their range. But of all these issues, intense development struck me as the most important threat to brookies in North Carolina. The trend I witnessed in Georgia, and then in Tennessee, also was evident in North Carolina.

To better understand the status of North Carolina's brook trout, I met one morning with Doug Besler, a state biologist and chair of the state's Eastern Brook Trout Joint Venture steering committee. When asked about the most immediate threat to brookies, Besler confirmed my suspicions: "Rampant development poses the most immediate threat to brook trout populations." We were in his office just outside Asheville, which seemed appropriate given the ubiquitous evidence of development in the form of new houses, golf courses, and resorts.

The Asheville area and western North Carolina have become a magnet for those who seek cool air and mountain scenery. While the region has been a vacation destination for more than century, since the 1990s a hyperexpansion of new homes and developments has occurred. The point is that a landscape can be loved to death. As is the case on the Tennessee side of Great Smoky Mountains National Park, brookies and myriad other creatures found in highland forests are being hemmed in by development on the park's North Carolina side.

That evening, I drove around Asheville and picked up brochures and read the advertisements for some developments in the area. As I examined their sales pitches, some were ironic. One project claimed to focus on "wellness and sustainability." A building company claimed its development provided "home settings and amenities [that]

emerge naturally from the land—as if they always have been [there]." Another hyped a home that "elevates the North Carolina mountain living experience to the peak of quality, beauty and enjoyment."

I wondered what local, old-time mountain folk or Indian peoples might think of highfalutin language about living in the mountains. And what did those descriptions really mean? I can appreciate the effort to create developments that fit into, or minimally impinge on, a landscape, but can luxury homes and all the infrastructure they entail truly appear to have "always been" part of a natural landscape? I could not help but think about wastewater treatment, runoff from hot paved surfaces, and the inevitable erosion from homes and roads built on the sides of mountains.

I again was reminded that intense development and nature are inherently incompatible, as are intense development and brook trout and their cold crystal-clear waters. While it seems obvious, it is hard to accept. We can push nature, but our demands eventually break natural processes and laws. That is not to say the folks buying those homes do not love the landscape. Most probably do. But rampant development comes with an environmental price: someone's home water loses out.

This breaking point was clear when I visited The Cliffs, a development with a golf course designed by Tiger Woods on a mountainside just outside Asheville that holds at least two brook trout streams. When I had been in Asheville previously, the golf course was still under construction, and the barren red clay slashes—future fairways—looked like open wounds. A lot of earth has to be moved to create a golf course on the side of a mountain. As I stood outside the gates, erosion from the barren red earth awaited the next thunderstorm. Sure, a few erosion-control barriers were in place, but you can't build at a large scale on mountains without the earth's hemorrhaging. Before I could enter the development to look for a stream, a security guard approached. It was time to move on.

Nature is messy, dirty, wild, and often inconvenient. That was clearly evident as I gazed up at Tiger's luxury development. We want nature until it gets in our way. At that point, something has to give, and that something, in this case, was trees, slope, soils, and trout.

As bleak as the future of North Carolina brookies might look on private land around Asheville, they are not in imminent danger simply as a result of development pressures. Hundreds of streams within Great Smoky Mountains National Park and the region's national forests contain brook trout. But I am quite sure the overall range of the brook trout will contract during the next several decades, perhaps significantly. Luxury housing developments, golf courses, brown and rainbow trout, all coupled with climate change, will increasingly limit brook trout to the highest elevations and most remote headwater streams, usually only those in protected areas. We are fortunate to have protected areas, but these high-elevation populations will also become more isolated from one another, as they are in Tennessee. Such a condition is unnatural.

Still groggy after another night of bluegrass music, I drove out of Asheville the morning after my real estate tour as I admired the muggy city and its urban amenities in a beautiful landscape. I suppose it is easy to criticize development, since I do not live in the area, and maybe I am even a little jealous, because the area has so much to offer. Changes I saw seemed to be happening at an astounding pace and scale. As more people are drawn to this city and its landscape, its only native trout are forced farther into the mountains and hollows. And as the brook trout retires deeper into the wilderness, so does a part of the region's soul.

CHAPTER 5

Virginia

IT WAS ONLY 7:00 a.m. and already too hot for coffee. The air was still and thick, and gnats buzzed around my ears. I was staying near Standardsville, on the eastern side of the Blue Ridge Mountains, at the home of Mason Overstreet, my trusty research assistant and University of Alabama student. Mason was the teaching assistant for my fly-fishing class in Tuscaloosa, so he was eager to show me some of the streams he grew up fishing, his home water. We loaded the car to prepare for our drive to Shenandoah National Park, about an hour to the west.

Under a large oak tree in a nearby pasture, cattle sought shelter from the sun. Mason and I were excited about the prospects of a day of fishing, but the summer had been hot and dry, and we suspected the stream would reflect that. I felt a bit like Don Quixote, accompanied by Sancho Panza, facing the weather windmills. My head was also a little heavy: Mason's parents, George and Susan, are quintessentially gracious southern hosts, which meant the previous night included a long dinner with excellent food and wine. As Mason discussed the day's options, we decided to first fish a stream in the higher reaches of the park, hoping

to find cooler water and hungry brookies. Mason assured me that, even on the hottest days, we could find plenty of fish high up in the cold headwater streams. These were his home waters, and he confidently promised me brookies.

Although I have been a frequent visitor, I have never spent long periods of time in Virginia. I usually drove through the beautiful Shenandoah Valley on my way to and from Maine on Interstate 81 when I made that annual pilgrimage. But even quick glimpses had left a favorable impression. When I first started making the trek during the late 1990s, small tobacco farms still dotted the rural landscape of southwestern Virginia. In late summer, gold-tinged tobacco leaves splashed color at the base of hillsides set against the background of the pine- and hardwood-clad Shenandoah Mountains. A dozen years or so later, I could rarely spot tobacco fields from the interstate.

With its historic towns, neat farms, expansive parks, and the lingering influence of Thomas Jefferson and James Madison, Virginia has always been one of my favorite states. Jefferson's vision of a nation dominated by yeoman farmers has always appealed to me because in it

I can see a modern path to local food sustainability. And his deep interest in natural history made his a significant scientific mind of his time. Our nation's history remains a part of even the modern Virginia landscape. Colonial Williamsburg, Monticello, and the vernacular architecture in random towns and hamlets across the state provide windows to the past.

Shenandoah National Park protects three hundred square miles of the Blue Ridge Mountains in the central Appalachians. While this area is not as biologically diverse as the southern Appalachians, the park provides habitat for an impressive array of creatures, including more than 200 bird species, more than 50 mammals, 51 reptiles and amphibians, 30 fish species, and more than 1,300 species of vascular plants. This diversity owes to not only the topography but also the park's mid-Atlantic location, which straddles both northern and southern Appalachia, resulting in a large mountainous ecotone (a transition between ecosystems). With its thousands of miles of brook trout waters and highland forests, the Blue Ridge and Allegheny Mountains area has biogeographic connections to both New England and the Deep South.

Visitors packed the park on our first day out. As we approached the entrance at Swift Run Gap, at least a dozen cars were ahead of us. Given that it was Sunday, I guess I should have expected this, but the traffic at the entrance was surprising for eight in the morning. I started to grumble and began to talk about Edward Abbey's similar disdain for "industrial tourism" at national parks in the West, but I realized we were in line in a large SUV like so many around us. On that day and in that line, I was part of industrial tourism. I guess others, too, were seeking refuge from the heat in the highlands. But we found no relief—the air was thick with late summer humidity. Despite that, the forest was bone dry after the long hot summer. Trails were dusty, and campfires had been banned. As we walked the trail, swatting at the buzzing mosquitoes, I found relief in the hope that I soon would be standing in a cool stream. I rarely ever see other people when fishing headwater streams, but today would provide no such seclusion. We would be casting in front of, and sometimes around, other tourists.

This was really the first time I had ever fished on a small brookie stream around so many people. Small groups hiking on streamside trails stopped to watch us fish. Sometimes strangers would shout questions over the roar of the stream. It felt strange, almost foreign. Here I not only fished but served as tour guide. I do not begrudge curious passersby, but they set a different tone for the day. There is little room for contemplation when a stranger is taking your picture. In one moment, I was casting to a nice pool I was certain would yield a nice fish, when a teenager scrambled around the rocks above the pool to get a better view just a feet from where I was standing. He was in both my physical and mental space. Needless to say, I caught nothing in that pool, which was adjacent to a trail; the brookies scrambled for cover when they saw his shadow. I did not expect solitary streams in a crowded national park, but I did not expect people would literally be looking over my shoulder.

Make no mistake about it: Fly-fishing is often a solitary, even selfish, sport. We anglers want pools and rivers to ourselves. Most of us cringe when we see a car at a roadside parking strip near a favorite stream or when we walk around the bend of a stream and find someone already fishing in a trout pool. Even when I fish with Mason, we almost always split up and fish different sections of the waterway. Fly angling is a closed meditative world. I am somewhat embarrassed to admit it, but the only time I willingly open up that world is when I fish with my sons, close friends, and students, because I so enjoy helping them learn the craft of fly-fishing. But that openness also has its limitations. When I am on a stream alone, casting to biting fish, I rarely miss my family or anything else in that moment. I have everything I need at that moment. Again, that may sound selfish, but my contentment while angling is such that I willingly accept that label.

Our eventual destination that day was White Oak Canyon Run, a small steep freestone stream with cold deep plunge pools throughout its course. The stream was steep enough in some sections that I wondered how brook trout navigate its course. White Oak Canyon is a well-known brookie stream in the heart of the park. Mason had fished this stream many times and thought it would be a good choice in hot weather, so we saved it for the afternoon. His judgment proved correct. After hiking a few miles down into the canyon, we turned around and started to fish back up the stream. As we moved higher, we caught a fish out of nearly every plunge pool. These were small fish, six to eight inches at most, but that was expected, given the size of the stream. They were also really colorful, with squash-yellow bellies and bright blue spots on their sides. It was August, but those brookies looked like late fall fish in spawning colors.

The day turned out to be a great one. Both of us caught and released a couple dozen brookies. Our success was representative of Virginia's extensive brook trout resources.

Highland Virginia has hundreds of similar streams. But it was not just the fish that made the day special; we saw a lot of wildlife on our way to and from the streams. A buck with cinnamon-colored velvet antlers grazed near our vehicle, seemingly oblivious to the cars and hikers around it. Mason, who was fishing some distance downstream from me, even saw a mother bear and two cubs. He was lucky she moved on after looking his way. Banded water snakes were also a familiar companion that day. Although these are not venomous snakes, the sight of any snake coiled in a brush pile inches from my hands still was disconcerting. These snakes had the same thing on their mind as I did: brook trout. They thrive when the water is low and brookies are more exposed. Because I have spent a lot of time in Louisiana's swamps, I assume, at least in the South, that any snake around water is a water moccasin. So I irrationally jumped back every time a water snake appeared. We also found a dead timber rattlesnake. The six-foot snake had been run over by a car on a busy park road, a casualty of industrial tourism. We counted twelve rattle segments, meaning it probably was an old snake.

Our success that day and the two days that followed was not especially surprising. This was not going to be a three-streams-or-bust kind of visit, as South Carolina had been. Of the 2,350 miles of wild trout resources identified by state fisheries officials, approximately 80 percent continue to hold brook trout (although most are not exclusively brookie streams). Highland Virginia has enough brookie streams that one can find hungry fish even during the hottest part of the summer.

All of Virginia's trout waters have been inventoried since 1980. This is a remarkable accomplishment, given the extent of trout waters, and shows the interest of both state and federal biologists and local citizens in the state's cold-water resources. That we know the location of all the state's trout streams is almost mind-boggling. It is hard to fathom that we know all the locations of anything in nature, let alone creatures found in thousands of miles of streams. This baseline information will become especially vital as the planet continues to warm and southern states lose brook trout streams. The inventory already has been used to show that Virginia has lost at least 38 percent of its brookie waters. According to Larry Mohn, a state biologist who has since retired, before the arrival of Europeans brookies occupied more than five thousand miles of streams in what is now the Commonwealth of Virginia. As is the case in so many other areas, much of this loss of stream mileage has resulted from agricultural practices, the stocking of nonnative trout, acid rain, and development within watersheds.

While the acidity of streams has declined in recent decades, thanks to passage of the Clean Air Act in 1970, some headwater streams are actually more vulnerable to acidification today because highland forests have experienced significant tree diebacks. The wooly adelgid, an invasive insect, and gypsy moths have caused extensive destruction, especially to hemlocks and oaks, which in turn affects a forest's ability to neutralize some of the problems caused by acid rain in headwater areas. Ninety-five percent of all hemlock within Shenandoah National Park are now dead. Skeletal trees, some large patches even, are now evident in much of Appalachia, and the geographic borders of national parks do not keep pests and pathogens out. This is especially troublesome in headwater areas,

because they have so little buffering capacity to begin with. So, in Shenandoah National Park and Virginia overall, air and water quality offer both good news and bad news. The good news, says Larry Mohn, is that "acid rain is still a problem, but it looks like we have turned the corner." The bad news, he adds, is that "forest cover and species composition is changing due to invasive pests such as the wooly adelgid, which allows the impacts of acid rain to linger on."

After talking with Mohn, I felt optimistic about brookies in Virginia, given the overall stream mileage with fish, even though a 38 percent decline is a significant loss. Too often, however, my optimism is cockeyed because I catch a lot of brook trout. It is easy to be lulled by an immediate experience. The bigger picture in this case is one of large declines in the number of brookie streams. Unless skeletal trees are visible in an area, the actual effects of acid rain and global warming are difficult to comprehend during a short visit in any landscape, hence my overly optimistic perception. Instead, brookies are experiencing an invisible poisoning, a slow death that seems implausible in a national park. Even though air quality standards have generally improved throughout the eastern United States, given its geography and elevation, Shenandoah continues to receive more acid rain than areas closer to sea level. That a threat, such as development or logging, is not immediately apparent does not mean threats do not exist. I need to remind myself of that while I fish in the middle of a seemingly pristine national park.

Although brookies still live in thousands of miles of streams in Virginia, biologists at state and federal fisheries, as well as local Trout Unlimited chapters, continue to work to restore habitat, remove nonnative trout from the park's streams, and reduce the effects of pollution. On our second morning in the park, Mason and I met with David Demarest, a park fisheries biologist who was conducting a stream survey with a team of seven volunteers from the Student Conservation Association. It was another hot morning, and, because they were electroshocking the stream (which temporarily stuns the fish, making it easier to survey them), everyone was wearing waders and rubber gloves, so it felt even hotter than the air temperature. The purpose of the survey was to catalog any and all fish species present and remove any nonnative trout found in the stream.

The survey stream was small because it was both a headwater stream and in need of rain. So I was a little skeptical about just how many brookies they would actually find. The stream had a few pools, but it was not exactly ideal trout habitat during the hottest time of the summer. I was surprised, however, as brook trout after brook trout floated to the surface after receiving a mild jolt from the battery pack. Most were small, a few inches long; many were the young of the year. But, along with tiny fish, a few relative giants in the eight-inch range surfaced. The group placed fish in holding pens until they could be measured and returned to the stream. Mason and I wished we had fished the stream before the survey crew showed up. Demarest just laughed at our surprise and told us, "You'd be surprised how many nice brookies we find in these small streams." When we saw them float to the surface, we immediately began peppering him with questions about other streams in the park that contained big brookies. Our minds had quickly turned from conservation science to angling.

Like their counterparts at Great Smoky Mountains National Park, biologists in Shenandoah—Demarest among them—have removed all nonnative trout from survey streams. National Park Service policy requires the restoration of ecosystems with native species. But, as in Great Smoky Mountains National Park, this policy has not been uncontroversial. Demarest is a moving target, because he is in the field conducting surveys all summer. He told me anglers often complain that removing "perfectly good trout" is a waste. Even he acknowledged, "I do hate to kill any trout. I have had really mixed emotions about taking out a nice brown, but brook trout are the park's only native trout, and as such they need streams where they aren't preyed upon by the big browns. It isn't feasible to remove all browns and rainbows from the park's streams, and, in reality, we'd have a battle on our hands if we tried. But it is important that we have enough brookie-only streams so that their future remains secure. They are an important part of the park's heritage."

His team found nothing but brookies that day, so this stream was in good shape. As they packed up and began to plan for the next sample location, Mason and I began planning which stream we should fish next. The rest of our day was going to be easy. But Demarest and his team would work their way through the park stream by stream all summer, no matter the heat, the bugs, or the hours. This mission and work ethic are why Virginia has produced a complete inventory of its trout streams. Good science and good data are the results of hard work.

Leaving Demarest and crew to continue their work, Mason and I decided to head up to the Rose, one of the park's better-known trout streams, which Chris Camuto writes about eloquently in *A Fly Fisherman's Blue Ridge* (1990). That day, the Rose was warm, apparently too warm for brookies. We fished each and every pool for several hours. I had nothing to show for my efforts, at least no trout. One pool I came across was representative of the day—a long deep run that had a dozen or so brook trout holding in the current. My eyes initially bulged with excitement, because one of the brookies was about ten inches long—big for this or any southern stream. I clearly could see the cream-tipped fins of the others in the gin-clear water. The biggest hovered near the head of the pool, where the constricted current entered. As I watched the fish hold near the bottom, a few caddis and stone flies flitted downstream past me.

It looked easy. I would cast my small caddis fly to the top of the pool, let it drift a couple feet, and enjoy watching that nice brookie come to the water's surface and sip my fly. But, of course, appearances can be deceiving. I threw the caddis probably a dozen times before I changed flies. Another dozen casts with a tiny caddis pattern resulted in the same thing: nothing. No movement from those fish at all. The largest brookie remained almost motionless in the current. I knew the water was low and warm, but these fish looked so ready to eat my fly. I moved on from that pool, leaving the fish in the same spot as when I arrived. As I walked around the pool and up the stream, I tried to convince myself those fish weren't brookies at all. I wanted to believe that they were suckers, thereby salving my ego. We were in the dog days of summer. I felt hot, tired, and depleted by the heat. It seemed like the fish felt the same way after a long summer. It was strange to catch so many brookies one day and none the next, but in addition to

being warmer, the Rose probably had been fished hard all summer. Those brookies had no doubt seen their share of Adamses, caddises, and woolly buggers.

Mason had slightly more luck, catching a couple of small brookies on his favorite fly, the Mr. Rapidan, a fly developed by the local fly-fishing guru Harry Murray. He constantly harasses me about using this fly, no matter where I fish. I've had minimal success, so it has become a running joke between us, with me disparaging his favorite. I prefer the big gaudy stimulators, but Mason seems to think these easy-to-see flies are too generic or too commonplace. I was happy he caught a few fish, because he had insisted that, since I was a guest, I should fish the best pools. I had not protested. I saw trout in many pools as I hiked up the stream. Usually they hovered close to the bottom, in the coolest water they could find. But even though I approached each pool slowly and quietly, the brookies ignored all my offerings.

I made few casts in the afternoon. Instead, I continued to hike and wade my way up the stream and then sat on the bank to watch trout in the remaining pools. I had no reason to bother the fish. They were not interested, and I could have worked my way through all my fly boxes without any brookies to show for the effort. Sometimes we have to realize we cannot force nature, in this case brook trout, to do what we want when we want. Desire does not mean catching brookies. The fish do not care that I have traveled hundreds of miles to be here, although on slow days I wish they would reward my effort.

During the heat of the afternoon, the forest was quiet. Other than running water, the only sounds came from squirrels turning over dry leaves on the forest floor as they buried acorns. Birds were silent, except for a distant woodpecker drumming high in the canopy. Throughout my travels, the sounds of insects, birds, and, of course, water usually overwhelmed me, but the landscape seemed parched and tired this late in the summer.

Late in the afternoon, the buzzing drone of cicadas finally began, reminding us that dusk was approaching and it was time to start our hike out. The brookies probably would have stirred a bit as the air cooled slightly late in the evening, but another grand dinner with the Overstreets was waiting for us.

Aside from catching brookies, my other goal for the week was to meet individuals who were working on brook trout conservation or restoration projects in the state. So the next day I met with John Ross, who then chaired the Virginia Council of Trout Unlimited and was nationally known in fly-fishing circles for his books, including *Trout Unlimited's Guide to America's 100 Best Trout Streams* (1999) and *Rivers of Restoration*: *Trout Unlimited's First 50 Years of Conservation* (2008). He also was one of the driving forces behind the formation of the Interstate-81 Coldwater Area Restoration Effort, or I-81 CARE, Virginia's original large-scale and long-term project to improve trout habitat at the watershed level all along I-81, including sections of the interstate in Tennessee and West Virginia.

In Virginia, I think of this interstate as the trout corridor, because it straddles the Blue Ridge Mountains (to the east) and Allegheny Mountains (to the west) along most of its path. Thus, it represents the home—at least historically—of brook trout in Virginia. Yet this corridor is also littered with streams that, as a result of agriculture and development, no longer support brookies.

Interstate 81 bisects the picturesque Shenandoah Valley, which historically has been agricultural, home to row crops, hundreds of thousands of cattle, and millions of poultry, especially turkey. I-81 CARE was formed in 2007 with the goal of mitigating some of agriculture's long-term effects on trout streams. Ross saw this program not as a quick fix with a defined checklist of streams to work on but as a long-term commitment that will grow big brook trout and improve overall cold-water habitat. "Some of these streams once grew big brookies, not just the small fish we associate today with our headwater streams," Ross told me. "There is no reason that, if we address long-term air and water quality issues, as well as improve stream habitat, some of these streams can't grow big brook trout again, especially the spring creeks."

Ross envisioned Virginia's becoming a destination for brook trout fly-fishing. "I want people to think of Virginia the way we think of Maine today" as an angling destination, he said, but he also acknowledged that achieving this goal presents many challenges. "We are talking about years, even decades, to make these changes. But we are fortunate in Virginia, because there is a great deal of interest in our cold-water resources." With the I-81 CARE framework, he was hoping that world-class brookie fishing will someday return to Virginia. "They [big brook trout] were here in the past," he said, "so if we make a generational commitment, there is no reason we can't have that type of fishery here again."

The past and present trout resources in the I-81 corridor, and in the Shenandoah Valley specifically, led Trout Unlimited to create the Shenandoah Headwaters Home Rivers Initiative. This effort evolved from the original I-81

CARE idea, given the interest in trout in the Shenandoah Valley and the number of TU members in the area. This effort was bolstered in 2021 with a $2.9 million grant for restoration. With more than four thousand members in Virginia, Trout Unlimited can get restoration projects done and exert political influence at the state level. So, based on membership and cold-water resources, it is natural that TU would invest in one of its core areas. Seth Coffman, TU's coordinator of the Shenandoah Headwaters Initiative, has a long list of projects, streams that once were home to wild brook trout populations but have suffered from agriculture, development, and any number of other human impacts. The region really presents a mixed bag in that plenty of streams still have thriving brookie populations, but many others on private land either have been degraded for many years or even generations or are threatened by the recent explosion of population and development.

Coffman has the tricky task of often working with multiple landowners when designing and proposing a restoration project. One such example is protecting the Dry River near Harrisonburg. Coffman calls Dry River one of the best brookie streams in the region outside Shenandoah National Park. But it also became a favorite destination for the off-road community, who drove their trucks through the main channel of the river and created deep mudholes. The City of Harrisonburg owns more than seven miles of the Dry River because it is part of its municipal drinking water supply. When Coffman and local TU members made city officials aware of the damage being done, the city posted the property and placed large boulders in specific entryways to keep the trucks out. Today Dry River remains one of the most popular and healthiest (in terms of number of fish) brook trout streams in Virginia.

Having a home waters coordinator is important, because Coffman serves as a guardian for trout. If a new development is proposed, he checks on its legality and its probable impact on nearby streams. If a new mining or logging operation is under consideration, he can mobilize concerned citizens and educate the public about environmental impacts. Perhaps most important of all, he can work with landowners interested in protecting or restoring brookie habitat. While plenty of brookie streams remain in the Shenandoah Mountains, their future is more secure if more brookie populations exist inside or outside the park.

The last stream we fished during my four-day stay in Virginia was the Rapidan. The stream is well known and regularly fished, partly because a road runs parallel to it, making for easy access, and partly because of its history and beauty. High up on the Rapidan sits President Herbert Hoover's old fishing camp in Shenandoah National Park; it was his retreat from the heat (perhaps literal and figurative) of Washington, D.C. Hoover even wrote a book with a conservation bent about his angling experiences, *Fishing for Fun—And to Wash Your Soul*. A few cabins still stand but all are locked. It is no major tourist destination, more of a curious stopping point for the many day trippers or anglers who fish up the river. Because the river is already fished hard, I appreciate that the park's administrators are not trying to draw crowds to the location, but fly-fishers should celebrate any U.S. president who fished for brook trout and wrote a book about it.

After our slow day on the Rose, Mason and I were not optimistic about the Rapidan. But its history made it a must-fish location. To continue our conversation about TU in Virginia, we met up with Urbie Nash in Waynesboro and caravaned up to the stream. Nash, an engineer by trade, is past president of the Shenandoah Valley Chapter of TU, past president of the Virginia Council of TU, and a former member of Trout Unlimited's national board of directors and its national executive committee. For decades he has been deeply involved in various trout restoration projects in Virginia, from sounding the alarm about the decline of Mossy Creek, Virginia's most famous trout stream, which wallowing cattle had largely destroyed by the 1970s, to working on a variety of contemporary restoration efforts in the Shenandoah Valley such as that for the South River.

Before splitting up to fish, we stood along the stream in bright sunshine and talked trout with Nash, a lean, friendly, and intensely focused man. "We've come a long way compared with twenty or thirty years ago, but the Shenandoah Valley still has some major problems," he said, adding that he was especially concerned about what cattle and poultry operations are doing to spring-fed creeks. He became animated, waving his arms and raising his voice while discussing nitrogen runoff from those operations. "The amount of runoff from these farms is equivalent to millions of people living in the Shenandoah Valley! If we think about the pollution in terms of people," Nash continued, "then it sounds more concrete. And we are talking about millions of people. But the fact that it is coming from these farms, and not people, doesn't make it any less of a threat to the area's streams—or wells, for that matter. Runoff, siltation, cattle in the streams, we have seen it for decades. Yes, in some ways it has gotten better, but there is still a lot of work to do." With that, we split up and hit the stream. It was already midmorning, so it was time to fish before the sun rose any higher and the water became warmer.

It was another hot day, with temperatures expected to reach the low 90s, so I welcomed the opportunity to wade in the deep pools. The Rapidan is a steep stream dominated by large plunge pools, so it is easy to sneak up on the next higher pool, which sometimes sits at eye level. Mason and I would sneak up to the next pool and peer over the top of the boulder, sometimes spotting brookies hovering almost motionless just on the other side. Other times, with the sun high in the sky, a brief shadow spooked the fish just as we approached. Sometimes we cast blind over the boulder with one of us directing the other about when to set the hook. This added a new level of challenge. Fishing was slow again, but each of us managed to catch a couple of small brookies. Given the size and depth of some of the plunge pools on the Rapidan, I would like to return in early summer. I imagine with cooler temperatures and more insect activity, it might be a spectacular river to fish.

I also caught an eight-inch chub. While under normal circumstances that would not be worth reporting, this was a large bluehead chub (*Nocomis leptocephalus*). This large minnow-like fish has a sleek silvery and scaly body; it is unremarkable except for the dozen or so small spikes atop its head. So while I was a bit disappointed by the slow trout fishing, the chub was fascinating in a strange and even beautiful way, the last thing I expected to pull out of a deep pool in Shenandoah National Park.

The spiky-headed chub made me contemplate good fish and bad, or trash, fish. When I was a kid, my dad would refer to carp or buffalo as trash fish. We never ate or kept those species, but they seemed to fight as hard as or harder than bass or catfish, the good fish of the day. I think most anglers probably have some sort of similar hierarchy, or, at least, they generally mentally segregate good and bad fish. I place brook trout above bass, rainbow trout, and certainly the chub, although Izaak Walton, in *The Compleat Angler* (1653), defended the chub, writing, "Well, scholar, you see what pains I have taken to recover the lost credit of the poor, despised chub." Yet the bluehead chub frankly fought as hard as those small brookies. I was certain it was a nice trout until I had it in my hand. I doubt I will become a chub fisher, but I think the surprise of finding a new species, one that has unusual features, forced me to reevaluate my fish hierarchy, at least for a few moments. Perhaps Mr. Walton would have been proud.

I got an early start the next day, headed up I-81. The predawn hours were still and silent, except for a distant mockingbird whose exuberant singing defied the hour. As I hummed along the darkened highway, I felt good about brook trout in Virginia. Other than Maine, no state has more public and private groups and individuals working on brook trout research or restoration efforts than Virginia. The combination of TU and state and federal biologists has created a powerful advocacy group for cold-water conservation.

Virginia has ambitious restoration goals, but these goals are based on a rich natural history and a willing volunteer workforce. Certainly, challenges exist, including the ongoing effects of climate change, which could force brook trout into the highest reaches, and continued pressures from agricultural runoff and development. But improving and restoring habitat in areas suitable for brook trout but degraded by human activities can counter those losses. That may sound like a fanciful wish, given that some of the threats, such as climate change, are beyond the control of local trout advocates. But I would not bet against them in Virginia. It may take a generation or two before we again see the monster brookies that once plied Virginia's spring-fed creeks, but anglers are patient.

CHAPTER 6

West Virginia

By entering West Virginia through the George Washington and Jefferson National Forest and then the Monongahela National Forest, I saw dense green highland forests, the beginnings of the rugged Allegheny Mountains, and plenty of freestone streams with trout. This was a big change in the landscape from the bucolic Shenandoah Valley of Virginia to the east. As I climbed higher, thick fog and mist cloaked the slick blacktop. Spruce, pine, and hemlock crowded the narrow highway, creating an eerily beautiful midmorning atmosphere. As I came around one tight corner, a ruffed grouse flew chest high over the road, just in front of my car.

While West Virginia has a not-undeserved reputation for removing mountaintops to reach its coal, this picture-postcard part of West Virginia is home instead to wilderness as deep and rugged as anywhere in the eastern United States. Large mammals, such as eastern elk and bison, survived here longer than anywhere else east of the Mississippi River, with John James Audubon noting the presence of a few remaining eastern elk in the Alleghany Mountains as late as 1851. During this trip in mid-May, I found myself surprised by the scenic landscape. However, I had entered the state on U.S. 311, which took me directly into the Jefferson National Forest, a section not affected by present-day coal mining.

My journey to find brook trout in West Virginia began in Slatyfork, a tiny settlement in the middle of Monongahela National Forest, in the east-central section of the state. I stayed at the Elk River Inn and Cabins, a lodge that sits near the headwaters of the Elk River and the heart of West Virginia trout country. The river's name sheds light on the region's rich natural history. The Elk in May is a good-sized limestone river, forty or fifty feet across in some areas, with some pools deep enough to swallow me, which made spring wading interesting, to say the least. At the time of my visit, the area had had a wet spring, so the river was high and cloudy. I initially planned to fish only for brook trout, but my guide, Gil Willis, owner of the Elk River Touring Center, suggested we spend a bit of time on the Elk to see if we could entice a few big browns from the deep holes. He at first could not believe that I'd stay virtually on the banks of the Elk and not want to fish it.

Although he was excited to look for brook trout, the Elk was too good to pass up, he told me.

Crossing a rushing river in spring is an unforgettable experience and not necessarily a pleasant one. Every step is measured, thought out. You can feel your feet becoming increasingly unsteady as you move into faster water. As an angler who spends most of his time in tiny streams, I rarely encounter such violent water. You test each rock for stability, because the river will send you swimming if you stand on one foot for a second too long. Will this rock move? Is it flat enough to step on? Is it covered with slippery algae? Walking almost anywhere else is an almost unconscious act. Wading a raging river requires you to keenly focus on the *here and now of every step*. Crossing an uneven slick bottom in fast-moving water might be as precarious as a toddler's first steps—you could topple at any moment. It's unnerving because of the danger, but it literally makes you connect with the landscape.

Fishing in the Elk shares some aspects of fishing any mountain stream. A big river has a lot more pockets to cover, and most are deeper and bigger than a mountain stream's, especially when the water is high. But the process of finding fish remains the same: you cover every fishy-looking pocket with a cast or two, slowly moving upstream as the fly floats back. On large rivers, however, I often find myself trying to reach the farthest pool or bank in the belief that the far water has more fish than nearby water. It's also fun to pick a distant target and stretch my arm. I sometimes convince myself that the farthest pool seems more pristine, fished less often by others, so it must have more fish. Not until a nice trout shoots out from under my feet do I have been careless and not mindful.

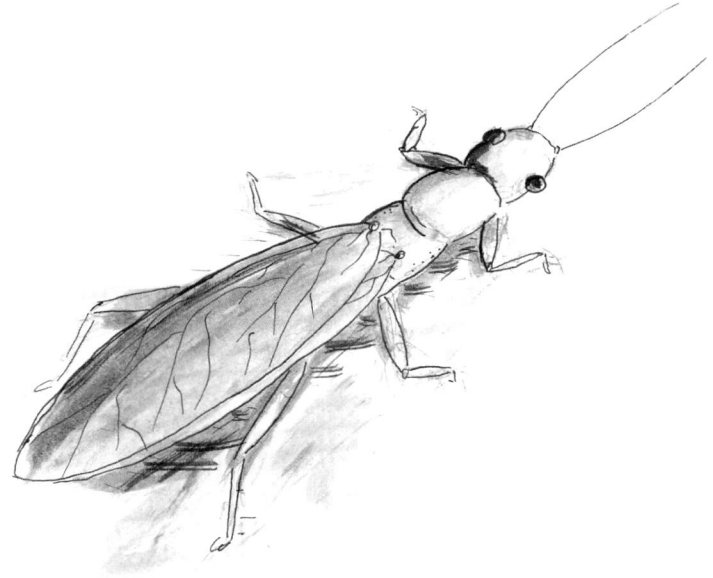

Stone fly nymph (*Isogenoides hansoni*).
Drawing by Frederick Steinberg.

After crossing the Elk and finding refuge in some slower shallow water, we fished upstream with a dry fly and nymph dropper. I cast and slowly stripped my line back, keeping my eye on the strike indicator. Even the slightest movement on the indicator could be a trout. I repeated this slow-motion pattern as I worked my way against the current. As I worked up river, the occasional large stone fly flew past and bounced its way downstream, which I took to be a good sign.

Willis's expectation was right on. After wading and casting upstream for about twenty minutes, I hooked a beautiful twenty-inch brown trout just a few yards away from me. It peeled line off my reel, as it headed downstream and was quickly into my fly line backing. (Backing is a tougher line attached to the lighter line on a reel; backing acts as an extender and insurance policy when a large fish decides to fight.) My rig that morning was more conducive to hooking brook trout, so I worried that I was not up to landing that angry brown trout. I was fishing with a light five-weight line and rod, so I kept telling myself, "Enjoy the fight, enjoy the fight, it may not last." I was mentally preparing to lose the fish. Up and down, back and forth across the river we went. The brown was not happy to be hooked. The current did not help my cause as I also had to deal with the weight of the swollen river. I have lost other big fish in similar situations when the current and fish put too much pressure on my line and small hook. When the fish became still after about fifteen minutes of fighting, I knew he had gone under a rock. My heart started beating even faster. I could feel the fish scraping against the bottom to try to free itself from that tiny nymph. But, lucky for me, the fish shot downstream again. After several more minutes, it began to tire and Willis was able to get below it in the river. He seemed as nervous as I was, urging me, "Keep the rod tip up, slow down, and don't force it." He knew catching a big fish like this was not an everyday occurrence. A few minutes later, he had the fish in the net, and I was able to admire the beautiful golden brown up close. The trout's belly was a dark copper color, with large black spots splattered along its back, a Winslow Homer piece of piscatorial art.

After a few photos, I held the fish underwater in the swift current for several minutes to help revive it. It had been a long fight on light line, so the fish was tired. I was tired. My right hand, which had been clamped to the rod for more than twenty-five minutes, quivered. With a sudden jerk, the brown bit down on my thumb, and with one thrust of its tail it was off into the heart of the river. I was left with a punctured thumb, though I would trade a minor injury for a trophy fish any day. That night, I had dinner at the lodge with a friendly group of anglers from Virginia. Word had spread that I had caught the big fish of the day. Catching a nice fish takes the edge off traveling alone; other anglers are always interested in good stories. While I prefer to fish alone, it is nice to share a meal with like-minded anglers after a good day.

Throughout my travels, I had a love-hate relationship with big browns. They are lovely fish. When hooked and rolling at the surface, showing their golden flanks, they remind me of fall leaves floating on the surface of an Appalachian stream. The sight of that brilliant roll on the end of my line is why I fish. All anglers receive a jolt of adrenalin when a fish—any fish—shows itself. But I also know those big browns eat a lot of smaller brook trout. The Elk River has many small tributaries that contain brookies, but when some of those fish migrate downstream and enter the Elk, as they occasionally do during the year, they become meals for the big browns. So while I will never intentionally break off or kill a brown, I do lament how they affect native fish in many streams wherever they have been introduced. Big fish must eat smaller fish to get that big. Even though they have a large effect on native fish populations, I have to admit they are beautiful and a hell of a lot of fun to catch.

After fulfilling the urge—or maybe even need—to target big fish in the Elk, Willis and I hiked up into the Monongahela National Forest to search for brookies during the late morning of my first full day of angling. The forest around Slatyfork has many small brook trout–laden tributaries that eventually feed into the Elk River. As we walked, we noticed that the forest floor was literally crawling with hundreds of red efts, a newt species common in this part of the country. I had to concentrate to avoid stepping on the bright orange creatures scampering through the mud and leaves in search of the water in which they would complete their life cycle.

We hiked for about thirty minutes high up into the national forest to a classic brookie stream. Small steep plunge pools filled with small fish. The water ran really high and fast because of a recent rain, so we spent little time trying to catch trout. After a few dozen casts without a strike, we chatted a bit about local environmental issues before heading down. These streams cascade under a canopy of mature forest well above the Elk River, even above decent-sized waterfalls that kept the rainbows and browns from reaching the brookie habitats. Although not catching any brookies was a disappointment, seeing the stream, forest, and newts was worth the time. While we hiked downslope, I heard the songs of migrating warblers—northern parulas—from the treetops. I had recognized the bird by its short high series of notes: *peep, peep, peep peep.*

Now that I had seen the mountain streams, Willis wanted to show me another brookie location. As we hiked back downslope, gingerly crossed the Elk again, and drove away, I was perplexed, because we were not heading deeper into the wilderness. I asked what he had in mind, and he responded sarcastically, "It's a secret location." We pulled off the main highway by a tiny creek that looked more like a drainage ditch than a trout stream. I had told Willis about my quest to catch brook trout along the entire Appalachian chain, so he had brought me to the nearest, most accessible brook trout stream around. We had a good laugh about this location, because brookies are so closely associated with rugged headwater streams high in the watershed. Yet here was a stream and a nearby beaver pond with an ample supply of wild fish. Brookies do like wilderness, but they can be found in other landscapes, even altered ones, if the water quality remains high. I admit, somewhat reluctantly, that I caught my first West Virginia brook trout in what looked like a roadside ditch.

Like those in the many states I visited during my travels, the headwater streams in the Monongahela National Forest of West Virginia at first appear to be largely protected from direct human assault. But even though they lie within the national forest and therefore are assured some degree of protection, factors beyond the control of national forest biologists, such as acid rain and climate change, can affect the forest's streams and trout. According to Chris Shockey, a Trout Unlimited council member and my TU contact in West Virginia, most of the headwater streams in the Monongahela suffer from acid rain, while development, poor land-use planning (or failure to enforce existing laws), and logging on private lands continue to isolate streams outside the national forest. By the end of my travels, I had seen the effects of these threats in many states.

In the Slatyfork area, ill-planned or even illegal development stood out as a significant problem for not just brook trout but for the larger environment. While searching for brookies, Willis pointed out an area in the upper Elk headwaters where several landowners had dug a new river channel in an effort to better control the flow of the river. Apparently, the river had flooded their property, so they dug a new channel that was deeper and straighter, complete with earthen levees.

It was a shocking sight: the headwater channel of a world-class trout stream obliterated by a bulldozer in someone's field, a field with cattle in it, which meant a stream with cattle in it. There are trout in the Elk, so the landowners' actions did not destroy the river, but creating a new channel through a barren and open cow pasture with no state oversight was hardly a benefit for the Elk River's ecosystem. The silting from the new channel alone surely had negatively affected the cold clean-running river. And now cattle waste finds its way into the stream. Also, with no trees protecting the new channel, water temperatures will become unpleasant or even deadly for trout in the warmer months.

The day after Willis showed me around the Slatyfork area and I caught my West Virginia brookie, I met with Todd Petty, a professor at West Virginia University in Morgantown who specializes in fisheries-related research. As an avid angler and outdoorsman with an appreciation of trout and their environment, Petty is anything but bookish. We drove up to and through the Snowshoe Mountain Resort in his pickup to fish and check out the Upper Shavers Fork, a trout stream that had been heavily affected by a wide range of human activities in the past but was making a comeback because the decline of the human population had given nature a reprieve.

Upper Shavers Fork is a high-elevation stream on Cheat Mountain. At more than 4,800 feet, Cheat is the highest point in West Virginia and home to a fascinating remnant alpine environment dominated by red spruce. The Cheat area is home to the largest spruce forest south of Maine, more than forty thousand acres. According to the Nature Conservancy, the Upper Shavers Fork watershed contains one of the largest concentrations of rare species in the central Appalachians, including the West Virginia flying squirrel, Cheat Mountain salamander, and the last viable population of the Cheat minnow. Breathing deeply high up on the mountain and surrounded by an alpine ecosystem, I felt as if I was in northern New England.

The high elevation is what attracted the developers of Snowshoe. The view from the top of Cheat Mountain is fantastic, but this large-scale development, opened in 1974, has had a significant impact on the surrounding trout streams and rare spruce forest. I am not accusing Snowshoe of doing anything illegal, but with development, especially development in environmentally sensitive areas, comes pollution. On the other hand, at 244 acres, the resort is relatively compact compared with developments around Great Smoky Mountains National Park. Snowshoe is one more example of the development dilemma. We want to visit unique and scenic areas, but how do we visit without setting in motion their eventual degradation and even destruction?

In addition to the alpine ecosystem, the ghost town of Spruce, on Cheat Mountain just northeast of Snowshoe, provided a fascinating backdrop for fishing. It was founded in 1902, but all that remains of Spruce are a few concrete foundations, mounds of coal tailings, and scattered heavy machine parts. Before its abandonment in 1926, Spruce was a logging town complete with a hotel, paper mill, and various other businesses to support guests and residents. Demand for spruce trees must once have been great, considering the trouble of laying railroad tracks and building an entire town in this cold and often snowy landscape.

Leaving the forest and entering the large open area where a thriving town once stood was surreal. I was surprised that Shavers Fork still held trout—imagine the refuse dumped into that stream for twenty-four years a century ago. Pulp mill waste, coal ash, human waste, animal carcasses—no one alive knows how rank this stream might have been. Trout are indeed sensitive to environmental changes, but they can reappear in a stream once it has had time and the opportunity to heal. That stream reminded me that nature is never static; it is always evolving in response to human-driven and natural changes.

We did not have a great deal of time that day on Shavers Fork, but I fished long enough to catch a few small brown trout from a slow-flowing part of the stream that passed right through the main part of the old town. While I was fishing, an antique black locomotive from the Cass Scenic Railroad State Park pulled up belching thick black smoke and emptied its few cars of tourists on the slope above us, providing an odd moment when we looked at them and they looked at us, each group wondering what the other was doing there. Given that we had taken only a few small rough back roads to get to the stream, I had not expected to see another person, let alone a group of tourists.

What Petty really wanted to show me was not the ghost town but the complete disconnection of the main stem of Shavers Fork and its many small tributaries. The rail line had cut the small brookie streams off well above the main stem, making it impossible for brook trout to move in and out of individual streams. Culverts had been built under the rail line with no effort to connect these passageways to the larger stream. As fish populations become isolated, they also can become less genetically fit and even disappear unless new fish arrive to add new genes to the pool. This connectivity or lack thereof is a major theme in Petty's research. His research in West Virginia found that the vast majority of culverts are partially or completely impassible for fish and other aquatic creatures.

This was clear the day we spent on the stream: culverts draining headwater streams had created small waterfalls high above Shavers Fork. Water flowed from the dingy gray pipes that formed the culverts and over steep rock piles placed to prevent erosion before it eventually made it into Shavers Fork. However, there were no channels, only large rocks, so brook trout had no way to move from one stream to another. Culverts might not be the first thing that comes to mind when thinking about threats to trout (and many other aquatic organisms), but they are a huge challenge to the long-term survival of small-stream brookies.

After spending three days in the Slatyfork area, I drove to the northeastern part of the state—the heart of the Allegheny Mountains—to meet Chris Shockey near Elkins. At the time, Shockey chaired West Virginia's Back the Brookie program, so he oversaw TU's brook trout restoration efforts. Today, he is the West Virginia representative on Trout Unlimited's National Leadership Council. Shockey, who is intense and has the lean physique of a distance runner, travels throughout the United States in search of native trout.

I set out for Elkins at dawn. It was a cold damp early May morning in 2012. After a few days at one base camp, I was restless to start heading north again. Although Elkins was only an hour and change north of Slatyfork, I was happy to inch closer to Maine while catching some bookies along the way during what was then my annual summer trip. I missed my family, but I also felt great anticipation about angling in a new brook trout stream later that morning with another brook trout–obsessed angler.

As I left Slatyfork, dew covered the newly plowed fields with sprouting corn, beans, and greens. It was cool enough that steam rose from manure piles in barnyards and from the roofs of still-darkened homes. I passed a well-preserved barn with a large Mail Pouch Tobacco advertisement painted on its side and was struck that this landscape probably hadn't changed much in decades. This was agricultural Appalachia at its finest. Most folks in the area were still involved, at least part time, with small-scale farming, timbering, hunting, and fishing. The landscape hinted at poverty, with abandoned cars and trailers scattered about, but it certainly wasn't the desperately poor Appalachian landscape so often depicted in the national media.

To safeguard its location, I call the stream we fished that day Deep Pool. According to Shockey, Deep Pool was literally dead in the 1980s, a casualty of acid rain. Through the hard work of the West Virginia TU since then, however, the stream had been reborn, mainly through placing crushed limestone in its headwater areas. Limestone buffers acids, making a stream more alkaline. This solution is not perfect—it does not address the acid rain at its source (coal-fired power plants in Ohio and the Midwest)—but dumping limestone into headwater streams has been a relatively simple and effective method to resurrect many of them. And given that the brook trout evolved in headwater streams with little buffering capacity, they have evolved to be the trout species most tolerant of acidic conditions. Adult brook trout can tolerate pH levels as low as 5.0, but acid mine drainage and acid rain often produce even lower pH levels. (For comparison, household vinegar's pH is about 2.5; 7 is neutral.)

Shockey and I started our day far below the headwaters and fished and hiked our way upstream a few miles. This was not your typical tiny brookie stream; it was actually wide enough and open enough to cast. Best of all, it was pocked by deep pools. Shockey suggested I bring weighted wooly buggers that could get down deep, but I honestly thought he was joking, because so few brookie streams have truly *deep* pools. But this one did, along with tannic water so dark it resembled a well-brewed cup of English tea. I initially insisted on using dry flies. It's not that I am a dry-fly purist, but I was fishing for brook trout, and, under most circumstances, brook trout love big juicy hairy dry flies. Given the deep dark pools of the stream in question, I should have heeded Shockey's advice from the outset. But I kept throwing big dry flies, which were not even getting a sniff, while he was catching beautiful brook trout after beautiful brook trout on his wooly bugger. I finally admitted defeat, switched to a simple wooly bugger, and caught and released brook trout. Plenty of them.

That stream was far from dead. Because of the tea-colored water, the fish were dark, some almost black. With orange bellies and black backs, they looked like Halloween pumpkin brookies. These fish were stunning, all the more so because Deep Pool had none just a few years ago.

Those fish and all the other creatures found in that stream demonstrated once more that nature can rebound from significant pollution. The stream also stands out in my mind as another shining example of the good works provided by groups such as Trout Unlimited. Somebody has to pay for, haul, and dump that limestone, and TU members always seem to be there.

The experience of catching brookies that day was all the more exciting because of the scenery, including huge boulders, sheer fifty-foot cliffs, the associated waterfalls, and an overarching mature forest. Along some stretches, I had to combine boulder climbing with fly-fishing. I pushed my cheap waders to their limits, as I climbed up and slid down the huge stones. Later in the morning, I eyed a largely hidden deep pool on the other side of some rounded boulders the size of pickup trucks. I knew a nice black-backed brookie had to be lying in that pool. If I wanted to reach it, I would need to climb those smooth rocks and jump to the far rock. Shockey thought I was a bit crazy, yelling out, "You must really want a brookie!" But I managed to scramble to the last boulder and, after organizing my fly line, made a perfect cast, catching a nice-sized dark brookie. The top of that rock was precarious. One slip and I might have ended up at the bottom of that pool with no fish and a busted rod, but it would have been worth it.

Of all the streams I fished during my travels from Georgia to Labrador, Deep Pool stands out as the most stunning. And if I had to choose a single stream to fish again in eastern North America, Deep Pool would be it. Next time, I will take a few days to fish its length and camp along its banks, probably solo. Shockey was a generous fishing partner who gave me plenty of pools in which to catch fish, but really observing a stream requires exploring it alone without conversation.

After fishing into the afternoon, we drove into the headwaters along U.S. Forest Service roads to see where the volunteers had placed the limestone in the stream. As the water flowed out of the spruce forest, it was even darker than it was downstream, where we had fished earlier in the day. Here in the headwaters, just off a forest service road, the TU members had dumped crushed limestone in piles along the stream. This seemingly simple practice has helped restore hundreds of miles of brook trout streams in West Virginia.

Despite the volunteers' efforts, I could not help feeling a little uneasy about the future of brookies in the Mountaineer State. The crushed limestone lining the banks of that stream was and remains its lifeline. Without it, the stream will be threatened again because acid rain has not gone away. The trout, insects, crawfish, salamanders, and anything else that lives in that water could disappear. I would like to think that liming those streams is just the first step in a process that ends with no acid rain, with no need for dramatic human intervention to rescue a mountain stream and its fish. But I also know this is naive. Policies, administrations, economics, and values change. Until we, as a society, make an even greater effort to wean ourselves from power sources dependent on cheap sulfur-laden coal, all we can offer those fish, bugs, and salamanders is a short-term and tenuous lifeline.

Hope and dismay seemed permanently yoked that day, and then I came upon a sight that gave me a tangible reason to think that permanent large-scale positive environmental change regarding our reliance on coal has begun: wind turbines, dozens of wind turbines, on high hilltops in

West Virginia. From a distance the spinning blades looked like a child's toy, but as Shockey and I drove closer, I saw how enormous these new sources of energy are. Even here, in King Coal country, in the land where mountains are literally and obscenely blown apart to get at that coal, stood a wind farm.

As night approached and my time in West Virginia waned, I felt deep satisfaction at having fished a beautiful restored stream full of hungry fish. Part of me wanted another day on Deep Brook, but I had accomplished my angling goals. It was time to continue north on I-81 through the Shenandoah Valley of Virginia and get closer to Maine.

PART II
The Mid-Atlantic

CHAPTER 7

Maryland

AS I CROSSED THE Potomac River at the West Virginia–Maryland border on Interstate 81, I left the South and entered the mid-Atlantic East, already a more northern-looking landscape. I was still in the province of the Allegheny Mountains, at least an eastern sliver of them, but I was in a new cultural landscape. Farms were larger and looked more affluent. Stone houses and bank barns began to dominate the landscape. No hardscrabble Appalachian structures, stone buildings speak of permanence. Although this shift truly begins in the Shenandoah Valley to the immediate south, stone buildings are dominant in Maryland. The settlers who moved stones from field to foundation had no intention of moving on. And, I noticed, they had treated their land differently, more gently, than had farmers on mountains to the south. I saw far fewer junk heaps and rusted cars hidden in the roadside forests.

At the time, I was still making my annual trip to Maine for the summer. And, as I passed through western Maryland on I-81, I always got excited, because the landscape reflects a true change in geography. The land looks different enough that I knew I had made progress on my long drive north. Although Maryland was a border state with competing allegiances during the Civil War and therefore shares some historical characteristics of the South, its modern incarnation sits squarely in the North.

Large numbers of German immigrants settled in this state beginning in the mid–eighteenth century and influenced this prosperous landscape. German-constructed stone farmhouses in Maryland and Pennsylvania remain family homes with manicured lawns and colorful flower gardens, providing lasting testimonials to people who valued permanence. Perhaps I felt attracted to western Maryland's landscape because it reminded me of favorite haunts in my home state of Missouri. During the nineteenth century, large numbers of Germans settled in Missouri's Rhineland, an area that falls roughly between St. Louis and Jefferson City, following the course of the Missouri River. Small red-brick towns such as Hermann are home to a number of wineries where Barb and I spent many an Oktoberfest afternoon. Although I feel as though I have lost my connection to my boyhood home, this brief pass through Maryland reminds me that I still carry important memories.

While Maryland's modern landscape is more prosperous than many points to the south, that does not mean its environment historically was managed with any more foresight than in surrounding states. In fact, since it was settled in 1607, Maryland has experienced widespread deforestation, mining, and large-scale and intensive farming, all of which resulted in soil erosion, water pollution, urbanization, and suburbanization.

Rural Maryland is relatively close to Washington, D.C., and Baltimore, and this part of the megalopolis—a region that extends from Boston to Richmond—is home to more than eight million people. Both Washington and Baltimore long ago began to suburbanize, leading to booming population growth in surrounding towns and counties. Rural Maryland and its cold-water ecosystems have felt the effects. Brook trout are absent from 62 percent of their historic range in Maryland. But their disappearance is actually not as extreme as in some states in the East, although it is not good news, either, given that the historic geographical range of brook trout in Maryland was limited. According to state fisheries documents, of the remaining 151 streams that are home to brook trout populations, more than half are in westernmost Garrett County, the least developed area of Maryland. Most of Maryland's brookie populations are classified as "greatly reduced," meaning that, within the subwatersheds where they occur, brook trout occupy only one to 10 percent of the area they claimed four centuries ago. They are literally backed into a corner in Maryland's panhandle.

Unlike their locations in southern states, including North Carolina and South Carolina, most of Maryland's brook trout streams are on private land. Only 11 percent of all brook trout streams are fully within state lands, so the vast majority of habitat is on private and a mix of private-public lands. This checkerboard pattern presents a number of potential problems. The most obvious is that an unprotected part of a stream can have an inordinate effect on a protected section, because water and fish do not recognize boundaries. Also, state fisheries officials, as well as Trout Unlimited (TU) volunteers, have a harder time accessing streams on private lands to initiate habitat improvement projects. TU volunteers throughout the eastern United States have told me that working with private landowners is difficult, if not impossible, because many do not want strangers on their land, even if the purpose is to create better fish habitat. If landowners are reluctant to permit TU or state authorities on their property, determining the health or number of brookies in specific streams is impossible.

Of the more immediate threats to Maryland's brook trout populations, suburbanization is the most serious. This was the first state I visited where extensive suburban development really threatened trout. Sure, local development posed problems around Asheville and Great Smoky Mountains National Park, but those examples were not connected to a major metropolitan area such as that around Baltimore. So, although I was happy to be moving northward toward my summer destination in Maine, I also was entering one of the most crowded regions in the United States, and crowds and brook trout don't mix well.

According to studies by state fisheries officials, brook trout populations disappear in watersheds where human land use exceeds 18 percent. If an impervious surface occupies more than 0.5 percent of the area of a watershed, the warm-water runoff from paved roads, parking lots, and rooftops typically will decimate the brook trout population. For example, a state study that monitored the status of brookies in six Maryland streams from 1990 to 2005 found no sign of the fish in three and attributed the losses to suburban development. And once a stream loses its brookies to development, getting them back is nearly impossible.

Pavement alone can be a death sentence. Think about the steam that rises off roadways, driveways, and rooftops after a summer thunderstorm. Hot water kills trout and other organisms dependent on cold water. Water that does not evaporate has to go somewhere, and, in much of the paved impervious East, that somewhere is into storm sewers that eventually send their water into streams.

While brook trout conservationists in Maryland face no shortage of challenges, the state also has an active and dedicated cadre working to protect the fish. The State Wildlife Action Plan of the Maryland Department of Natural Resources lists brook trout as a "species of greatest conservation need." Concern about the status of the brook trout prompted the DNR's Inland Fisheries Management Division of its Fisheries Service to develop the Maryland Brook Trout Fisheries Management Plan. The goal, the report says, is to "restore and maintain healthy brook trout populations in Maryland's freshwater streams and provide long-term social and economic benefits from a recreational fishery." Maryland has been one of the most aggressive states in protecting brookie populations on its public lands.

As I contemplated Maryland's policies, I felt a strong urge to get back to fishing, to stand in a stream and smell the surrounding forest. I also needed to check Maryland off my list of "caught-fish states," so my thoughts quickly turned to finding a brook trout stream. May had been wet in Maryland, and the rain continued to fall steadily during the first day and night of my four-day stay. The slate-colored skies, constant drizzle, and cold temperatures made me feel like I was going back a season, as it felt more like winter in Alabama, but I did not have the luxury of waiting for the weather to clear up. I had other meetings with trout in other states to the north.

On my first morning, I fished the Elk Lick Run near my cabin at the Savage River Lodge in Frostburg while keeping an eye on the sky and hoping for a break in the showers, but I had no luck. The water was high and muddy, so muddy I was never quite sure how deep it was. As I stood on the bank, I had a déjà vu experience: I was back at my childhood catfish creeks in the Midwest. My fishing was tenuous and awkward. I barely entered the stream because I had no idea how deep it was. It was unnerving. On a clear running trout stream, the bottom is visible, so moving across a stream isn't particularly mysterious—treacherous at times, sure, but not really mysterious. Because of the muddy water I felt really limited as to where I could go and how to approach the water. Jumping blindly into a potentially deep stream while in waders is not advisable.

The prospects of constant rain and high water were frustrating. I was determined to catch a brookie in each state I visited, and far western Maryland was not a location I could easily revisit, even when I was married and still making my annual trips to Maine, because most brookie streams are in the westernmost sliver of the state, a good distance from I-81. Although I did not catch any fish on my first day, I was impressed with Maryland's efforts to protect brookies in the Savage River watershed, where I was staying. That first day, in Savage River State Forest, a streamside kiosk held a sign informing anglers that "no-take" regulations were in place for brookies. Maryland is the only state in all my travels that imposed such a policy in an effort to protect its brook trout population. This policy took effect in 2007 to aid a study of the impact of the regulation on brook trout populations, but it has not been without controversy. Although I release all my brook trout, some regard taking a few fish home at the end of the day as important as fishing itself. I met one such individual the next day. I'll call him Bill. He was walking a surly miniature dachshund when he stopped to talk about fishing while I pulled on my waders and organized my gear. When I asked about the no-take policy, Bill said it was bullshit, because he had fished for brookies all his life and saw no reason for the ban on taking fish. He frowned and quickly moved on after I expressed my enthusiasm for the policy.

By the second day, I had realized that the best way to beat the rain was to climb into the headwater streams. Higher streams have smaller watersheds and are therefore less affected by storms. Alan Heft, Maryland's state brook trout biologist, had suggested in an email that I focus my fishing efforts on some of the protected tributaries of the Savage River such as Big Run and Poplar Lick in the more than fifty thousand acres of Savage River State Forest. But even there the water was off-color. So I hiked higher and higher to try to get to the clearest water. The forest was typical of an Appalachian mixed forest in early summer, leafed out and tinged with yellowish tree pollen that floated on

the surface of the numerous puddles. I continued about a mile upstream to get away from the road. I saw plenty of good-looking trout water along the way, but I didn't want to hear cars or see another person. When I couldn't pass up any more prime trout pools, I decided to fish—muddy water be damned. Chris Shockey back in West Virginia had told me that he had experienced fifty-fish days on these tributaries, so I had been really excited before the several days of rain left the landscape muddy and saturated. I had lowered my bar from expecting a big day to just wanting to catch a few brookies.

In the Mid-Atlantic, the Savage River area is known for its trout fishing, but the river itself gets most of the attention, because the area below the large dam at the southern end of the state park is under special regulations to ensure the brown trout are allowed to grow big. Native brookies, especially those in the Savage's smaller tributaries, are plentiful but are not trophies compared with the big browns in the main river. This meant I had a stream all to myself. A worn path followed much of the stream, but I saw no other fishers, no beer cans, no worm containers. I had driven in on a paved road and had seen plenty of cars in the park itself, but this was like being on a remote New England stream. Walking through several large orb spider webs was unpleasant, but it was a guarantee that the area had seen no foot traffic in at least a few days. These solitary days spoiled me and made the larger crowds I found on streams farther north almost unbearable.

I was not confident about using dry flies, given the high and dirty water, but even with the cloudy water, I quickly caught a few fish on weighted wooly buggers. Because I had prepared myself for another fishless day, I was pleasantly surprised when the first fish struck my streamer. Surprise fish are the best fish because I talk myself out of expecting any trout when conditions are poor. I reminded myself again of the Second Noble Truth: Desire causes suffering. Surprise fish make me feel as though I am either really lucky to run into hungry fish or a really good angler. I usually come down on the side of luck.

A different watershed means slightly different-looking fish. Compared with the tannic-inspired charcoal brookies I'd caught just to the south in West Virginia, the Maryland brookies I caught in the Savage's tributary were much lighter colored, even slightly dull. But even a dull brookie is a beautiful fish. Given that these streams normally flowed crystal clear, it made sense that the fish were lighter in color. The sides of the brookies were almost pewter with faint spots, but the fish still had the telltale orange fins edged with cream. The variations reminded me how important it is to protect not just species but populations within species, because they are genetically unique, adapted to a specific local environment. Brook trout, as a species, do not face extinction in the eastern United States, but if Maryland loses its brookie population, it will lose an important part of its genetic puzzle.

The next day, I set out to explore more streams within the Savage River watershed. I fished as many streams as daylight allowed, because all the rain had cost me valuable time on the water. The sun came out midmorning, and by early afternoon the smaller higher streams were beginning to clear. The sun's rays danced through the pale green leaves, reminding me it really was early summer. The sunlight made me feel more settled and comfortable, that somehow everything had been made right with nature. I

took the sunshine as a sign to take a break and snack on an apple that had been in my fly vest for a week.

I climbed a rise to get above the sound of the stream for a few minutes and watch the water and forest. It seemed like weeks since I had felt that natural warmth, and I welcomed it heartily. As if called on by some higher power, birds—mainly warblers, I thought—seemed to raise the volume of their early summer singing as the sun made its appearance. A male rose-breasted grosbeak called and preened in the branches high above my head. Its slightly pinkish-crimson breast stood out against the early season leaves. At that moment, I knew that I was not the only being in that forest who welcomed the sunshine. I reminded myself to be mindful when I fished, but all too often that mindfulness is focused exclusively on the water and the fish hidden in the shadows. When that happens, I often miss opportunities to enjoy nature's beauty.

Since I am always focused on fly-angling, I spend far too little time just sitting and watching the water and woods. When fish are hungry, I will often go all day without stopping for much or any food, drink, or contemplation. While I don't regret fishing hard and being focused, seeing that grosbeak in the canopy reminded me that I probably had missed a great deal of the forest's details while focusing too singularly on angling. As I sat above the stream, I suddenly felt selfish, because I was noticing in nature only what *I* wanted to see. Yes, I was there to fish, but fishing was not the only reason I chose to write a book about brook trout.

All the tributaries I fished during my time in Maryland were no-take, and it showed in the number and size of the fish I caught. The fish were large (for a small stream), and once the water cleared, they appeared in large numbers.

Signs about the no-take regulation were posted at every parking area and roadside parking slot. Maryland is serious about protecting its brookies. What was most remarkable was that I fished and actually caught them in a few streams that flowed right through or behind campgrounds. These were not seldom-used wilderness campgrounds; they were right off roads, so well used that the grass was worn down or worn away altogether, exposing the compacted mud around the fire pits. I suspected those fish were under a lot of pressure, so I fished in the campgrounds to see if the no-take regulations meant I would find fish in high-traffic areas.

I usually avoid fishing near campgrounds because of the crowds. Fly-fishing is a time of singular focus and reflection. I have no need to interact with others while on a stream. So when I pulled into an empty campsite in New German State Park, fished, and caught a few nice brookies on big dry flies, I was ecstatic.

The brookies were impressive, a few in the twelve-inch range. If fish were in evidence near a campground, Maryland's no-take policy had to be working well. Instead of finding their way to the skillet at eight inches, those fish had time to grow. And, according to Alan Heft, that is exactly what the no-take policy is intended to do: protect Maryland's limited brookie populations and create a better fishery by allowing fish to get bigger. Heft calls the Savage River and its tributaries biologically rich systems; they are not the typical high-elevation, highly acidic watersheds with little aquatic insect diversity typical of many Appalachian streams. So, if catch-and-release fishing is practiced, brookies can grow to a respectable size in this specific environment.

Before the day was over, I had caught a few dozen brookies, mostly on woolies early in the day, then dry flies later. As the streams cleared, the fish became more active and aggressive, hitting anything resembling an insect. Just about every pool and slack water held a fish. I always get a thrill while watching my fly slowly drift into a slack pool, only to be hammered a moment later by a hungry brookie. Fly-fishing, especially using dry flies, allows an angler to draw the depths of the stream to the surface. It provides us with a view into a largely unknown world. This is especially true in water that is muddy or even slightly off-color, because the stream appears to be even more mysterious. I felt fortunate to have success in Maryland, success that provided me with a window on an aquatic world that I would have otherwise not seen. Because so many dead Japanese beetles were stuck to my car's grill, I tied a big iridescent imitation beetle on my line. Their splash often scares small trout, but these fish were so aggressive that even the gaudy beetle worked.

That afternoon and evening were productive enough that I hated to see the long shadows as the sun faded. A late mayfly hatch, drawing fish farther out of the depths, made it even harder to leave. When fish are biting, I stay on the water as late as possible, but at some point I could barely see the fly and knew I had a ways to walk down and through the stream to find my car. My neck ached from my overweight fly vest, a sure sign I had been on the stream for a solid day of fishing. I had earned that night's beer and bar food at the Savage River Lodge. As I a shuffled to my car, legs stiff from standing all day, I heard the ethereal call of a whip-poor-will in the distance. It made me smile, because I associate the sound with summer evenings when I was a kid. My mood had changed dramatically in forty-eight hours, from trepidation about the weather to sheer happiness that I'd had the opportunity to catch fish on a stream in a forest disturbed only by the evening chorus of a distant whip-poor-will.

While I was happy about my success and Maryland's no-take policy, my small window of fishing success didn't mean the brook trout population in Maryland was healthy overall. It is easy to be lulled by a day or two of success, but to see the larger picture I met Alan Heft at his office in Frostburg on the fourth day of my visit. So I could better understand how pollution from the past affects the present-day landscape, he directed me to a nearby stream that remains heavily affected by acid mine drainage, the outflow of acidic water from coal mines—usually abandoned—that sometimes occurs after large-scale disturbances to areas containing sulfide minerals. Surface and ground water interact with disturbed areas and the sulfide to produce an often highly acidic brew, one that usually is too toxic for trout and other aquatic organisms. According to Heft, "We're still dealing with the aftereffects of mines that were closed years ago. It's a big problem for our cold-water environments, and it's tough and expensive to clean up."

Before our meeting, I thought I had found big brookie paradise in western Maryland; I was enamored with the no-take policies, and I had caught a lot of nice fish, so I naively assumed all was well in that part of the state. But my conversation with Heft enlightened me. I had not realized how coal mining had altered, and in some cases devastated, western Maryland's landscape. Following Heft's directions, I pulled off a roadway just outside Frostburg, and what awaited me was surreal: the headwaters of Braddock

March Brown mayfly (*Maccaffertium vicarium*).
Drawing by Frederick Steinberg.

Run, a rust-colored spring welling up from a long-abandoned mine. Streams should not be orange, yet there I stood, almost in disbelief, watching an orange fountain emanate from the earth. Western Maryland was heavily mined during much of the nineteenth century through the mid-1950s; as these mines were abandoned, many began to leach acidic water from their bowels. Heft pointed out, "You don't need to travel to West Virginia to see the impacts of coal mining. We have them right here in western Maryland."

This problem is not limited to this stream or even to western Maryland. Acid mine drainage remains a significant problem wherever mining occurred. Resource managers in coal-mining areas, whether active or defunct, must now lime headwater streams or use water dozers—essentially, elaborate water filters—to raise the pH of runoff to minimize the effects of acid. Here, as with the problem of with acid rain in Virginia, West Virginia, North Carolina, and Tennessee, these approaches deal with the problem after the fact, so they are nothing more than Band-Aids. A few cost-effective practical alternatives exist, but once a mine has been abandoned without following proper restoration procedures (or was abandoned before regulations were in place), expensive, invasive, and potentially dangerous mitigation efforts probably will not be feasible. Aside from the potential costs, dealing with acid mine drainage at its source—in the depths of an old unstable mine—is a daunting engineering challenge. So fisheries personnel do what they can to give brookies a fighting chance by addressing the disaster as best they can.

As I watched thousands of gallons of water that looked like liquid rust pour into the stream, I was struck by how decisions from the past can haunt even a quaint and prosperous-looking landscape. Nice trout in state parks and idyllic countryside scenes can easily lull you into a sense of security about nature. But the ecological realities of a region that has been settled, farmed, and mined for hundreds of years offer no escape. As I stood on the roadside that day while cars rushed by the nearby toxic fountain, I wondered what we are doing today, right now, that will lead future environmentalists to shake their heads on the side of a road. Will it be ill-planned development, global warming, or a problem related to fracking? We know how and why brookies disappear, but the hard part is incorporating that knowledge in thoughtful decision making today so that future generations will continue to have inspiring fly-fishing experiences in such places as western Maryland.

I did not want to leave Maryland with the weather clearing and the big brookies rising to dry flies, but Pennsylvania was waiting. As I drove east on I-68 to get back to I-81, I contemplated the brookies I had seen during the past several days. Even after witnessing the worst-case scenario that last day, I remained optimistic about the future of brookies in western Maryland, because the state, recognizing the value of its natural history and recreational offerings, has put in place some strict management regulations. This has protected some excellent brookie water, including in the Savage River watershed. But my mood was tempered by what is happening outside western Maryland's panhandle. In the eastern two-thirds of the state, brookies are probably on their way out for good. I do not like to accept that forecast, but that is reality along much of the eastern seaboard, which simply has too many people and too many paved surfaces.

George F. Thompson quotes what the late great geographer Eugene Cotton Mather (1918–99) famously wrote: "Clearly, we can make two observations about landscape in America—first, the original source of geographical knowledge is in the field and, second, this is going to be a great country once we get it all paved." That last thought seemed especially ironic as I zoomed along the interstate, rushing to get to my next destination along with the tens of thousands of other cars and trucks that daily ply this highway.

CHAPTER 8

Pennsylvania

I officially entered the North when I crossed the Mason-Dixon Line on Interstate 81 at the Maryland-Pennsylvania border just north of Hagerstown. I was firmly in northern trout country.

Pennsylvania conservatively estimates that it has more than five thousand miles of streams where brook trout are known to exist, but, as in the other mid-Atlantic states I visited, Pennsylvania's fly anglers and cold-water conservationists continue to pay for the environmental sins of their industrial ancestors. Coal was king in Pennsylvania, and, where coal is found, so is a trail of trout destruction, including approximately twenty-five hundred miles of flowing water affected by acid mine drainage, especially in the west-central part of the state. Beyond coal, hydrofracturing, or fracking, for natural gas represents a relatively new and potentially even more ominous threat to Pennsylvania's cold-water resources. Fracking requires pumping a great deal of water and a mixture of chemicals deep into the ground under high pressure to force the gas from layers of shale. Natural gas exploration and drilling companies have rushed to stake claims, many of which are in prime trout country.

I began my trout exploration in Chambersburg, a tidy town with a densely settled central core area with the feel of an old mill town. Settled in 1730, Chambersburg and its hinterlands witnessed most of the conflicts that took place on American soil, including fighting during the French and Indian Wars and the Revolutionary War and culminating with the torching of the town by retreating Confederates following their pivotal loss at Gettysburg in July 1863. This deep history is one aspect of Pennsylvania I find intriguing: the landscape, towns, and farms just feel old. Like Maryland, Pennsylvania has the stone farmhouses and bank barns that make both states appear settled and permanent. Amish farms, more stone barns, and densely settled old towns provide a glimpse of American history through their landscape.

My first stop in Pennsylvania was to fish the area's frigid spring creeks. Spring creeks, called chalk creeks in Great Britain, have a special allure within the fly angling world and have developed a dedicated following among many anglers. These crystal-clear waters flow from limestone bedrock, so they are rich nonacidic environments with a lot of aquatic vegetation, which means a lot of aquatic insects. For example, while fishing in Falling Springs Branch (just east of Chambersburg) on my second day, I cupped my hands and pulled up a handful of muck and watercress. Within the debris were hundreds of cress bugs that squirmed and scurried around my cold hands. Because of this rich assortment of bugs, spring-fed creeks often grow big trout, and big trout attract fly anglers dreaming of big fish. These big spring-creek trout are often the wiliest of fish. They have plenty of insects to eat, so only the most realistic presentations of flies and perfect casts will entice these brookies. For success, fishers often need tiny flies and extralong and extralight leaders. Spring creeks also are appealing because they maintain a consistent water temperature. The underground aquifers from which the water flows are less affected by seasonal temperatures and can be fished year round—in the heat of summer and the depths of winter.

Early on my second morning, I met with Mike Heck, who had been a Chambersburg-area guide for nearly two decades and continues to work as a guide today. Heck, who appears clean-cut enough to work in a bank, makes his living on spring creeks and knows them well enough to have written a book about spring-creek fishing tactics. He was one of my favorite guides during my trip north, because he did not just deliver his client to a stream; instead, he was genuinely enthused about teaching his clients how to approach and fish challenging streams. I wanted to continue on my brookie quest, but first I wanted to check out one of the local creeks and catch one of the big rainbows or browns that long ago replaced native brookies. I was eager to try spring-creek fishing and learn as much as I could from Heck about local trout conservation issues.

We targeted Falling Springs Branch, a well-known local creek. And, when I say local, I mean it flowed under I-81 near my hotel. When I fish with a guide, the usual routine is an early morning pickup and a long drive to the fishing destination. Not in this case—Falling Springs was right outside of town, in the middle of the Chambersburg suburbs, running right past suburban ranch houses. I have fished on ranches in the West but not in the yards of ranch-style homes. Not that I am complaining—it was an attractive landscape with the stream meandering near manicured lawns and gardens. Like everything I had seen in Pennsylvania thus far, the area around Falling Springs was settled and well worn but in an attractive, substantial manner.

Swaying weeping willows and blooming tulips—both nonnatives—brightened a gray morning. We pulled into the driveway of a private home and parked. I was wondering what we were doing until I noticed that the homeowner had graciously created two parking spots for the public, designated by a small attractive sign: PARKING FOR FLY FISHERMEN ONLY. Where were the NO TRESPASSING signs? Drive up, park, walk ten feet, and fish. *How easy*, I thought. I wouldn't be climbing steep and rugged freestone streams or tearing up my waders and bloodying my shins while bouldering up mountainsides. I would not need to slink under a barbed wire fence. As I fished, a lawnmower hummed in the background, slowing, cutting, then stalling, then cutting its way again through tall wet grass. The smell of a freshly cut lawn provided a strangely domestic but pleasant backdrop, as did the robins that searched for insects among the grass clippings.

Heck and I fished for a few hours with mixed results. We caught a couple of smallish rainbows, but I neither saw nor hooked any of the giant spring-creek trout. No matter, the creek and day were pleasant. The wide stream offered open easy casting, and the watercress-lined banks and crystal-clear water were a far cry from the philodendron-shrouded tunnels I often fished in the Appalachian Mountains, the same tunnels where I have left many flies in tangled branches. While no longer a brookie stream, Falling Springs has been the target of a great deal of interest and intervention by local Trout Unlimited members. Channel and riparian projects, along with efforts to limit streamside development, have been at the top of conservation agendas since 2000. Money donated by Trout Unlimited (TU) and Orvis has established a stream greenway to control development in particularly productive areas. This was no small effort, considering the value of streamside suburban property.

Of course, part of the allure of this stream is its suburban access and general openness, which create significant challenges for not only trout but the aquatic ecosystem overall. Where there are lawns, gardens, and pavement, there are chemicals, warm-water runoff, and road salts. Fortunately for the creatures that call Falling Springs home, the limestone geology provides a rich geochemical foundation that buffers the creek from some of these threats. That and a dedicated group of local trout anglers. The development pressures on Falling Springs were representative of what trout streams face throughout the East—especially in the more densely settled Northeast. It would not be the last time I fished a stream as cars whizzed past my back cast.

According to Heck, the fishing and conditions during my visit were consistent because of the habitat improvement projects undertaken since 2002, including the development of the greenway. The water and fish quality of Falling Springs had started to decline during the 1970s, but even with the improving conditions of recent years, the amazing mayfly hatches that drew fly anglers to Falling Springs in the eighteenth and nineteenth centuries have become more sporadic. And while rainbows in the twenty-inch range still cruise these waters, they are less common and extremely wary, given the natural abundance of food and the fishing pressure. When I fail to catch a trophy or a fish big enough to turn into a trophy, I have no trouble finding excuses for what probably amounts to my lack of skill. Catching large trout on tiny flies and tiny leaders is not something with which I have a great deal of success. Perhaps that is why brookies, with their outrageous propensity for slamming large flies, have a special place in my heart.

While Falling Springs has come a long way since the 1970s, general concern remains about development and stream conditions. As Heck and I chatted, we wondered aloud if a world-class trout stream, even one with a greenway buffer, can support big fish as land use intensifies around it. According to Heck, TU has worked hard to serve as a development watchdog, but a body of water surrounded by pavement, lawns, farms, and industry requires a delicate balancing act. For example, in September 2009 a small oil spill from a local factory in the Chambersburg suburbs found its way into upper Falling Springs. Although the spill was small and quickly contained, it demonstrates the challenges that urban and suburban streams face. Falling Springs has been blessed by its geology, but even limestone streams can be ruined.

After fishing Falling Springs in the morning, we headed out of town and into the countryside in search of a few

Grey drake mayfly (*Siphlonurus alternatus*).
Drawing by Frederick Steinberg.

brookies. We entered Pennsylvania Dutch country—grand Pennsylvania bank barns and farmhouses, with quilts and pies for sale, a true celebration of Americana. So much vernacular architecture in rural America has been replaced by prefabricated metal barns and trailers that seeing two-hundred-year-old structures was like traveling back in time. Each stone was shaped and placed with purpose; there is nothing random in these graceful structures. This rural landscape had been cared for, even cherished.

Like most guides, Heck cautioned that the brookies were small. As I had explained to other guides, I said that was OK because I was on a quest. As we drove, he pointed out various small streams, some of which still held trout, while others did not as a result of often-radical land-use changes such as the replacement of forest with pasture. We headed about ten miles east from Chambersburg toward Caledonia State Park, a heavily used state park with large parking areas, wide-open short trails, and plenty of rusting barbecue pits for weekend parties. Heck had not promised wilderness, and this was not wildland, but it was a clean pleasing park with a sweet little brookie stream shaded by still-thriving hemlocks. We simply pulled over, walked ten feet, and fished. It was not long before I hooked a bunch of small brookies on big bushy stimulators.

Though fishing next to a paved road while surrounded by picnic areas is not an experience I crave, and I do not normally hire a guide to drive me to a state park, the trip showed me the sort of waters to which brookies are consigned in this part of south-central Pennsylvania. A century ago, brookies—and big ones at that—would have been cruising Falling Springs. Brook trout, not nonnative rainbows, and browns would have been growing fat on the amazing mayfly hatches. And while a few brookies still live in Pennsylvania's spring-fed creeks, they are no longer at the top of the food chain. They have mostly been driven into small ecologically marginal headwater streams where an eight-inch fish is one to be celebrated. This problem is not limited to Pennsylvania, but it is more difficult to witness here because brook trout that once dominated those spring creeks were rivaled in size only by brookies found in northern New England.

After spending the night in Chambersburg, I ventured out to nearby Big Spring, another nationally known spring-fed creek. More recently it was known as a stream for a trophy brookies, whereas Falling Springs's native fish had declined a century earlier. Many in the eastern fly-fishing community consider this stream the finest brook trout stream in the East. Fly anglers such as Theodore Gordon raved about the size and numbers at the turn of the twentieth century. Charlie Fox wrote in *Rising Trout* (1978) that Big Spring produced phenomenal numbers of brookies. Nineteenth-century journals recorded daily catches of more than six hundred fish, some exceeding four pounds. I am not sure that any trout stream on the planet can produce those types of numbers today.

Big Spring is flows through fields and farms. It is easy to get lost in thought while sprawled on the grassy bank, enjoying the company of hundreds of bees getting drunk on clover nectar, but its present-day beauty and tranquility hide a damaged past. In this case, the harm came from fish hatcheries, which seems oxymoronic as hatcheries make rather than kill fish. The first hatchery was built on Big Spring in 1953 and became fully operational two years later. Regulars on Big Spring soon noticed that both insects and native brook trout were disappearing downstream from the hatchery. Hatcheries hold hundreds of thousands of fish, producing an enormous amount of waste. In the case of the Green Spring Trout Hatchery on Big Spring, its full capacity was 300,000 fish. It was the obvious culprit, because brook trout upstream continued to thrive. In 1958, an electrofishing survey found clear evidence of this relationship: While it found 641 trout per acre above the hatchery, it counted only 31 trout per acre below it, an astounding 95 percent reduction in trout biomass.

Big Spring got a reprieve in 1968 when the hatchery closed, and its brook trout populations below the hatchery seemed to rebound. But the Pennsylvania Fish and Boat Commission opened another hatchery on Big Spring in 1972. The irony is almost too great to comprehend. Instead of addressing the causes of the decline of the creek head-on, the commission decided to do what so many other states do: simply throw more trout at the problem. The new hatchery of course again overloaded Big Spring with nutrient-rich effluent from the 900,000 fish grown there annually. The native brook trout population subsequently crashed. Water-quality studies were conducted to determine how much the hatchery's effluent was increasing the nutrient load in Big Spring. Samples drawn from above and below the hatchery showed that nitrite and phosphate levels had increased by more than 200 percent, total phosphorous by 550 percent, and suspended solid wastes by 450 percent. Armed with these data, groups of anglers and local residents forced the hatchery to close in 1997. While Big Spring has made another comeback, the original strain of big brookies that inhabited the creek is likely gone forever, the result of pollution and stocking nonlocal strains and nonnative species. The glory days of this finest of brookie streams are unlikely to ever come around again.

The Pennsylvania Fish and Boat Commission was not alone in these practices. When native fish disappeared during the nineteenth and twentieth centuries, the typical response of most states was to build hatcheries and dump fish in their former habitat. Most eastern states, especially Pennsylvania, still abide by this philosophy. It seems that, so long as the crowds can catch a few dull-colored, ten-inch stockers, few people, other than TU members and other cold-water conservationists, really question, protest, or even remember that wild native trout once thrived and swam in a particular stream.

The morning I visited Big Spring, too many anglers were working the upper section of the creek, known as "the ditch," so I drove downstream to fish among the reeds, willows, and fields. I found plenty of great-looking water away from the crowds, although reaching it sometimes involved wading through knee-deep mud and bankside brambles. As I entered the stream, I sank into centuries of silt from nearby fields. I nervously hoped my next step would not leave me waist deep and unable to move. Clouds of fine sediment trailed me as I headed toward trout water, turning the clear water into a coffee-colored catfish river.

I was amazed that I did not see another angler on that stretch of water. Maybe they knew something I didn't. Maybe they had seen a buddy sink into the mud, with just the tip of an angler's fly rod identifying their pal's final resting place. The water, at least from a distance, was too trouty to ignore. They hid in pockets here, in undercut banks, under overhanging bushes and trees, under downed trees. They were out there; I just needed to get to them.

I started in an area recommended by Bill Ferris, a local TU member and one of the more active advocates for the restoration of Big Spring. I had met Ferris early in the morning to get a brief tour of Big Spring before striking out on my own. As we drove past it, he even pointed out a specific spot where a huge brown trout holds court. After a dozen casts and no luck with that particular fish, I slowly worked my way upstream, casting to any appropriate fishy-looking location. As I carefully walked along

the banks, barn swallows and swifts flew above the river, chattering and chasing bugs. This was a good omen, for birds and trout often share similar appetites.

Just upstream, more than fifty Canada geese had congregated near a streamside home. Under most circumstances, I admire these handsome birds and enjoy their company, but these became louder and angrier as I moved closer. And every time they spooked, their honking, agitated swimming, and occasional takeoff created a ruckus that no doubt sent any trout swimming for cover. I tried to approach them calmly and slowly, but that seemed not to matter. I finally just walked past them in an attempt to get quickly to the water above them, and my sudden move sent many skyward in an explosion of splashing, honking, and wing beats from the lawn and stream.

When I finally was able to concentrate on trout, I spotted what makes these creeks famous: two thick bright rainbows were holding in the current, occasionally taking bugs at and below the surface. By then it was late morning, and the sun was high, so as they moved to and fro to sip a bug, their red gill plates visible in the early summer sun. Standing in a stream, preparing to cast to a fifteen-inch trout, is both exhilarating and intimidating. Before you make a cast, you envision what you want to happen: the perfect cast, the perfect take, and the perfect photo. But actually making that cast, getting the perfect drift and presentation, is where fantasy meets reality. I believe that great fly anglers, probably like great major league hitters, do not think about the specific mechanics of that critical cast. They just do it. Overplanning and overthinking usually get me into trouble. I have had the most success when I see a rise, I cast, and, voila, it works out. The mental and physical come together in a succession of unconscious actions in the right place and at the right time.

As these thoughts churned in my mind, the small bunch of geese I had scared downstream circled around and flew over me; I could hear the wind whistling hard through their wings. Their shadows sent the trout torpedoing in all directions. I was left in midcast as I was about to deliver a perfect one. The line dropped dead, limp, on the stream. I almost could believe those geese had sought revenge.

No matter—more promising trout water was waiting. I slowly waded upstream, just on the edge of the bank so as not to become entangled in the flowing and swaying water plants. Their long green hair-like leaves waved back and forth just under the surface, like hula dancers in slow motion. I was standing knee deep in the creek, contemplating my next cast, when a snapping turtle the size of a small hubcap swam within a few feet of my legs. As it calmly moved past, I made my way upstream. I was surprised to see a snapper in this cold water, but that speaks to their adaptability. As I approached a section of the creek with a deep undercut bank laced with willow roots, I saw another example of a husky spring-creek rainbow, this one even larger than the previous two. It was one of the big old fish—pushing twenty inches—for which this stream is famous. Its silver flanks shimmered, as it held steady on the edge of the willow roots. This would be a tricky cast: sidearm, under the willows, so the fly could flow toward the trout. The current rushed into the roots, and I had to be mindful not to let the fly get carried too close and get hung up in the roots if the fish didn't take it.

I watched the fish for a few minutes while glancing up at the sky for any wayward sabotaging geese. The fish wasn't going anywhere, so I decided to take my time and add to my leader a few more feet of tippet material (7x), lightweight line that becomes nearly invisible in the water. As I glanced at the fish again, I questioned whether I should have used such a light tippet, but these fish are so spooky I did not think I had a shot otherwise. *Slow down, breathe, relax*, I told myself. I wanted to take no chances with this fish. I had to be perfect. After slowly but nervously adding more line, I delivered my cast. It was not a long cast, just a delicate one. I landed it almost perfectly, and my tiny blue-winged olive dry fly drifted just a few inches to the port side of the big fish, clearly within its feeding line. With a brush of its tail, the fish moved just to its right and took the fly. I set the hook, and the fish immediately dived back under the mass of willow roots and into a black hole. A few seconds later, the weightless line told me the bad news: the fish had broken the line in the roots. For a brief moment—well, maybe for several minutes—I was angry. I lost my focus on the present, on the stream and other trout, and sulked. A trout in the twenty-inch range is a big fish, especially if you have been catching small brookies for weeks. But after collecting myself, I realized I had done everything right and so had the fish. No hard feelings.

In *The Snow Leopard* (1978), Peter Matthiessen contemplates his failure to see that special animal during an expedition to Nepal. In Zen fashion, he confronts his desire head-on and asks himself: "Have you seen the snow leopard? No! Isn't that wonderful?" In trying to confront my own desire, I asked the same: *Did I catch the big rainbow? No! Isn't it wonderful? Aarrgghh.*

After a few days on the spring creeks, I headed to Clearfield in west-central Pennsylvania to meet with Ken Undercoffer, who was and is Pennsylvania's state coordinator for the Eastern Brook Trout Joint Venture and deeply involved in all things related to trout in the state. If there is ever a Brook Trout Hall of Fame, he will be a first-ballot inductee. Pennsylvania overall has a large and dedicated number of TU members and volunteers who have done innumerable good things for trout in the Keystone State. Falling Springs and Big Spring would not be in the shape they are in today without the intervention of TU. But TU members have given blood, sweat, and probably a good many tears to countless other streams, most unknown nationally. Their work has included stabilizing banks, enhancing instream habitat, reducing sources of stream sedimentation that affect water quality, liming streams degraded by acid mine drainage (AMD), restoring wetlands, collecting trash around streams and rivers, and educating the public and school groups about the importance of brook trout and their habitat. And leading the charge in these efforts has been Ken Undercoffer.

We met at his house, a small tidy ranch home in Clearfield, where he loaded me down with newsletters, published studies, and technical reports. (If text deals with trout in Pennsylvania, he has a copy.) He briefed me on how the Pennsylvania TU works on small-scale, individual stream projects while also participating in larger multiagency efforts to mitigate the effects of that mine drainage on watersheds. This is a difficult challenge but a critical one because poor mining practices of the past continue to harm so many streams in Pennsylvania. For example, the Catawissa Creek Project is reducing the effects of AMD in a 152-square-mile watershed. Catawissa Creek is a large stream, similar to a stream out West, that could,

absent AMD, naturally produce big brookies, Undercoffer told me. The Catawissa Project has established both active and passive mechanisms for treating water. In one effort, engineers have diverted tainted acidic water to limestone tanks, then through limestone tunnels to raise the pH. While the project was in its early stages when I was there in 2012, the results had been positive. According to Undercoffer, in one of the treatment areas pH had risen from 5.0 to 7.0 since late 2009, and in another area, pH had risen from 4.8 to 7.0 since 2001. This was significant, because any stream with a pH of 5.0 or lower is generally considered to be marginal at best and lethal for brook trout in some circumstances.

To Undercoffer, this project, which continues today, is important because reducing AMD and toxic metals from these streams rehabilitates miles and miles of prime brook trout territory. And not just tiny biologically sterile headwater streams but streams big and rich enough to allow brookies to reclaim a dominant place within the aquatic ecosystem. As Undercoffer excitedly told me, "These bigger streams can grow bigger brookies. This project could end up restoring some of the best brookie water in Pennsylvania."

Undercoffer wanted to show me Trout Run, because it is located practically next door to a recently closed coal mine, just a few miles outside Clearfield. As we drove in, we passed through an area that had obviously been strip-mined during the last year or two. An idle bulldozer waited for a new assignment. Robins and eastern meadowlarks paraded in the lush new grass—a huge lawn, really—that covered the area where an eastern hardwood forest once had stood, giving the landscape an unnatural appearance

room for optimism remains because many of the remaining brookie streams are on protected lands in the beautiful northern section of the state.

Like its surrounding states, New Jersey has a long history of trout angling and fly-fishing, and its stocking efforts date to the latter half of the nineteenth century, largely in response to environmental problems caused by rapid industrialization. So early stocking efforts were intended to mask the environmental degradation by dumping thousands of nonnative fish in degraded streams. However, in a strange twist and trade-off, New Jersey's long history of trout fishing has given it a large cold-water and sporting conservation community today, including ten Trout Unlimited chapters that supply both money and labor for various educational and restoration efforts.

I was heading to Hackettstown to meet and fish with Brian Cowden, who served as Trout Unlimited's Musconetcong Home River Initiative Project coordinator at the time of my visit and now is a partner in a river restoration consulting firm. As a home river coordinator, Cowden was TU's point man on all things Musconetcong, locally known as the Musky. "All things" in this case refers to coordinating volunteer efforts, working with local landowners or businesses that might affect the river, and planning restoration activities. Cowden was an enthusiastic advocate for trout in New Jersey and happily agreed to serve as my guide for a day to help me catch my New Jersey brook trout.

I began my tour on the Musky, a revered trout river in New Jersey and tributary of the Delaware River. Unlike many rivers in the region, the Musky saw relatively minimal industrial development along its banks and in some stretches resembles a meandering river found in an idyllic New England landscape. That does not mean, however, that the river is pristine or surrounded by unbroken wilderness. Few locations anywhere in the eastern United States can be considered pristine or wild, and this part of New Jersey is no exception. Even the Musky, some of which has been designated a wild and scenic river, is under pressure, especially from development and its associated pollution along and near its banks.

As we parked and then walked a short distance to the river, cars buzzed past on the road immediately behind us, reminding me that civilization was close at hand. Never mind the noise, though; Cowden wanted to show me a couple of the projects the local TU chapter had worked on in conjunction with other groups, such as the Musconetcong Watershed Association, to improve aquatic habitat and to help restore natural fluvial processes. These efforts included willow reforestation along eroding banks, dredging to eliminate some of the channel uniformity and silting, and removing two dams that no longer served a purpose but limited the movement of fish and altered the flow of water. All these efforts were impressive, given their cost and the amount of labor needed to complete them. Most of these efforts were focused on the river south of Hackettstown, considered the most scenic and best-protected stretch.

As I reflect on my trout travels, the Musky stands out because surrounding populations have strongly embraced the river. It will always be a high-use river with much human impact. Yet local people have embraced it, working hard to protect it and helping to restore it. New Jersey's long history of interest and participation in trout angling

Blue quill mayfly (*Paraleptophlebia adoptiva*).
Drawing by Frederick Steinberg.

has translated into present-day activism through TU in support of native and wild fisheries. As I've said, I prefer to fish on rivers without cars speeding by and occasionally honking in the background. But as Cowden and I climbed down the bank and slipped into the river to fish, a sense of calm came over me, and I was able to leave behind the discussion of the history of humans on the river and focus on the immediate task of reading the river for signs of fish.

The Musky did not disappoint. Brown trout were rising to sip mayflies in a deep ribbon of water along the far bank. As the morning's heat and humidity rose, mist hung just above the river. We took turns casting to the rising fish, and both of us caught a couple in short order.

What the Musky lacks in pristine stretches of river it makes up for in accessibility for the public, and that in turn generates interest in the major river restoration and conservation efforts. The widespread practice of stocking

nonnative fish (which New Jersey ended in 1990) has been criticized—and rightly so—in areas where native fish still spawn, but in altered high-use rivers stocked fish can connect people with their local environment. The Musky is a living testament to this. Sure, it would be great if the Musky still contained only native brookies, but those days are long gone and will never return. So I do not begrudge those browns, because turning back the ecological clock on a river the size of the Musky is utterly impossible.

After we caught the browns, Cowden suggested we move on in search of brook trout in the northwestern corner of the state. I am always reluctant to leave feeding trout, but we had more ground and water to cover. As we drove, Cowden described some of the state projects that focused on brookies. Perhaps the most impressive ones he described included Cresskill Brook in eastern Jersey near New York City and Mason's Run, which runs through a south Jersey golf course opposite Philadelphia. Both were amazing projects, because the streams still contained remnant brookie populations, despite the intensive widespread landscape changes around them and their proximity to what, for trout, are urban wastelands. They can hang on in tiny streams, so long as water quality and water temperature remain at survivable levels. Streams that run amid urban or suburban sprawl usually breach the water quality and temperature thresholds, and brook trout disappear. Mason's Run is an especially impressive restoration project because it involved completely rebuilding its channel and the intensive replanting of sections of the spring-fed stream. As is usually the case, TU contributed both money and labor. While neither stream is an angling destination, given their channel sizes and small fish, every trout stream in such a densely settled state is worth fighting for. These projects again demonstrate that trout anglers, especially TU members, are connected and committed to larger conservation and habitat restoration issues.

As we reached our destination in a protected northwestern part of the state (Cowden asked me not to name the location because these small streams are accessible to many, many anglers and cannot withstand much fishing pressure), I got the immediate sense that Cowden is a brookie man. Like me, he enjoys bushwhacking through the forest to look for small trickles with native fish. Sure, we both enjoy big fish on big water, but there is something pure about finding an unnamed stream with native fish.

After we parked on what appeared to be an overgrown logging road, we headed into a forested area in search of a first-order trickle that held tiny ancient-strain brookies. New Jersey was conducting a genetic survey of all its brook trout streams to identify heritage brookies, and the one we were searching for was in that category. After a thirty-minute walk through a dry crunchy forest while swatting at mosquitoes buzzing my ears, the stream appeared through a tangle of vines. Because it was late summer, the water was low and the brookies would be confined to the small deep pools. In early afternoon, the hot forest was quiet, the birds apparently waiting for cooler evening temperatures to begin their choruses. I was not greedy and did not want to stress the fish in the warm temperatures and low water, but selfishly I also wanted my New Jersey brook trout.

Luckily, the fish quickly cooperated. I caught a few small brook trout on bushy stimulators in the first two pools I fished. And, with that, my New Jersey quest was over. I could relax. With the heat high and water low, we decided to move on and explore the area a bit more, looking for slightly bigger water where fish would not be as stressed. We headed to a couple of smallish tributaries of the giant Delaware, but by then it was midafternoon, sunny, and hot, so we could not entice our brookie friends out of their deep cold holes.

We decided to call it a day. Cowden had a full slate of meetings the next day, and I needed to get on the road again. Although my time in New Jersey was brief, I was glad I had met Cowden and had gotten to see, firsthand, all the great work Trout Unlimited supported. No, the Garden State might not compare with Maine or Virginia in terms of angling opportunities for brook trout, but neither is it a trout wasteland. The Garden State lived up to its moniker. And, yes, I was glad I caught those New Jersey brook trout.

CHAPTER 10

New York

As I drove through northeastern Pennsylvania and entered New York, my mind raced with anticipation. I was heading to Roscoe, one of the oldest, most storied trout towns in the entire eastern United States. I was not exactly sure why, but I felt intimidated. Maybe it was my midwestern roots—anything New York seems exotic, foreign, and big. Or maybe I found it hard to imagine myself fishing the famed waters around Roscoe, the same waters fished by so many of the authors and founders of American fly-fishing, among them, Art Flick, Theodore Gordon, Sparse Grey Hackle, Ernest Schwiebert, Nick Lyons, and Lee and Joan Wulff. It was too late to turn back, though, as I had made reservations and booked guides. The Beaverkill River, Willowemoc Creek, and Ausable River were on my dance card, even if I was unworthy of carrying Gordon's empty trout creel.

Roscoe is in the Catskills, just a few hours north of New York City. The Roscoe area has been a trout destination for centuries, which means stocked browns and rainbows long ago replaced brook trout as the dominant fish. As in so many other states, stocking large numbers of nonnative trout continues. Brook trout are what drew early fly anglers here during the nineteenth century, but overfishing and massive environmental change led to the brookies' rapid decline. So the state turned to the more tolerant brown and rainbow trout. I had no delusions about catching many brookies, much less big ones, but I had to visit the Roscoe area for a few days because of its angling history.

I kept getting lost on the twisting, turning roads and started to wonder how Roscoe became so famous if finding it is so hard. And I was surprised that the rural areas around Roscoe seemed depressed. Many buildings and homes were ragged, like something you'd see in southern Appalachia. The countryside and even the town seemed to be in a long-term decline because of the departure of industries that once supported the region: traditional mill industries, logging, and tourism. Given the lack of any industry and major employers, the area today is more a weekend destination for residents of metropolitan New York. The decline of heavy polluting industries such as tanneries helped give rise to Roscoe as a center for trout fishing, so the area has traded industrial jobs for service jobs. Good for trout, bad for wages and benefits.

As I drove around a bend, before me lay the Beaverkill, locally known as the Kill, and its striking valley. It was late in the day, so the fading sun created a distinct contrast on the valley's walls with one side still bright, illuminating the greens of the early summer forest and affording its creatures a few more moments of a splendid day. The other side was dark, cold, almost black. The wide quiet river also was in shadows and looked deep, almost foreboding, more like a big western river.

As I entered Roscoe, I knew I was in Trout Town U.S.A. It has all the amenities a fly-fisher could want: plenty of fly shops, outfitters, small restaurants, taverns, small hotels and inns, and the nearby Catskill Fly Fishing Center and Museum. I had booked a trip with Catskill Flies, one of the better-known fly shops in town, so I thought I would check in before it closed. As I entered the store and introduced myself, Dennis Skarka, the owner, said from behind a counter, "Oh, yeah, you're the guy from Alabama. Why don't you have two first names like everyone else in the South?" I assured him that not only did I have one name, but we even wear shoes and have paved roads down in Ol' Alabama. Yes, indeed, I was in Yankee country. No matter, Skarka proved to have great information about local brook trout streams.

During the next few days, I spent plenty of time and money in that shop. And while flies and gadgets come and go, I bought several March Brown dry flies (an imitation of the early-season mayfly) that have proved to be my favorite fly pattern. This large fly, tied by Skarka, has cinnamon-colored hackle and a bright cream-colored post for better

visibility; it's easy to fish because it sits high on the water. Beyond its fish-catching ability, it is simply handsome. Flies really are works of art, and I laugh as I think about the first ones I tied as a kid, made out yarn and model airplane glue, but even those crude imposters caught sunfish. As I wrote this chapter, I was down to the last March Brown I purchased in Roscoe, so I needed to call Skarka and order more. I have never consistently caught so many trout on one fly as that March Brown pattern from Roscoe.

That evening, after checking into my room at the Reynolds House Hotel, a clapboard guesthouse built in 1902, I walked to a nearby tavern that turned out to be one of those classic small-town watering holes patronized by entire families for dinner and beers. I sat at a corner barstool and watched as extended families flowed through for chicken wings, hamburgers, and pitchers of beer. The scene reminded me of certain taverns back in St. Louis that serve as community centers. As the night wore on, the voices of patrons grew louder as they downed more beer. Children scurried around the bar and dining room as Heart, the Steve Miller Band, and Lynyrd Skynyrd blasted away on the jukebox. The atmosphere was almost that of a picnic, with people celebrating the early summer after an especially long, cold, dark winter. A local man, who was standing next to me at the bar with his wife and two kids while waiting for a table, even insisted on buying me a shot of bourbon to welcome me to town. I appreciated his gesture, because eating and drinking alone becomes a lonely business. He was impressed that I had driven all the way from Alabama to fish Roscoe's streams. I believe the question went something like this: "Alabama? What the hell are you doing all the way up here?"

The next day, after downing a lot of coffee to help clear the tavern-induced cobwebs, I set out for one of the nearby brookie streams. On Skarka's recommendation, I was going to fish a small tributary of the Beaverkill just a short drive from Roscoe. The stream was no more than ten feet wide in most places. Like so many other brookie streams I have fished, the routine consisted of climbing and fishing from pool to pool. This run required no long, majestic casts.

While it was close to town, this stream had the feel of wilderness after I had gotten away from the road. Once I started climbing, I never saw another person. Hemlocks towered over me, keeping the forest floor dark and damp—so damp that toadstools—an outrageous orange and cream-colored species—grew scattered on wet rotting trunks. Boulders, some the size of vw Beetles, were covered with wet lime-green moss, making them soft and life-like. The forest appeared surreal, or maybe I was just feeling a little surreal after the previous night of many beers, much cigarette smoke, and a shot of bourbon. During the early part of the day, I heard and briefly caught glimpses of a pileated woodpecker, its raucous call, machine-gun hammering, and brilliant red crest helping me pinpoint its location in the canopy.

This pleasant stream had typically small colorful brookies that were as friendly as the locals, quite willing to take my fly. Up and away I went, over boulders, through pools, under and over downed timber. Six hours later, with tired knees and bruised shins from slipping on slick rocks, I had caught thirty fish, maybe even a few more. All were small, six inches tops, and I released all of them alive. Like any angler, I find great satisfaction in catching fish, but equally satisfying is knowing I can return another day and find fish

because I did not drain the pool. Most of these streams were small and so barren of insects and cover that I couldn't imagine killing a brookie that had survived for three or four years against such long odds. Bad karma, bad energy. Releasing trout largely unharmed is my way of showing respect for nature. With a quick flick of my hemostats, I release the trout without taking it out of the water. I take joy in the catching, but I kill nothing. Native trout are as much a part of the landscape as old hemlocks and huge boulders, things that to me represent the soul of a landscape.

After my first full day, I placed New York in the "caught brookies" column. It had been a long day during which I had felt no need to eat and barely any desire to even stop for a drink of water. I was too focused on climbing, casting, and catching my way up the mountain, deep in that almost trance-like state that arrives when I am so focused on fly-fishing. I reach that state almost exclusively when the fish are active, as they keep my senses focused on the stream and not on my stomach. But after I stop fishing, hunger and thirst immediately kick in. A good day by all accounts.

Still, I could not help but be bothered that brookies, the only salmonid native to this area, were confined to those tiny pools, while the grand water below, the Beaverkill, was the domain of voracious German browns. These waterways are connected, and some brookies farther downslope might even enter the Kill from time to time. But the big water is dangerous for the natives. In the big water, these brookies are food for the big browns.

The origins of these present-day conditions date to the nineteenth century. Ed Van Put, in his excellent book *The Beaverkill* (1996), meticulously details the events that led to the decline of the brookie and the expansion of nonnative trout. According to Van Put, overfishing, logging, and industrial pollution all played key roles in this transformation. More specifically, beginning in the mid–nineteenth century, tanneries and acid factories generated huge amounts of pollution that degraded the region's water quality. From the late nineteenth century to the mid–twentieth century, fish kills regularly were reported on the Beaverkill and other area rivers with factories on their banks. Some kills involved thousands of fish at a time. Today it is hard to imagine that the beautiful Beaverkill ever could have been the vehicle of such destruction. Also, the Catskills became connected to New York City by rail in 1870, which led to an overfishing frenzy.

In reaction to the downward spiral of brook trout numbers, the State of New York implemented farsighted restrictions on taking brook trout. It even declared a no-take zone on the Middle Mongaup River in 1882. But because of public pressure and threats directed at wardens, they rarely enforced the regulations, setting the stage for the brook trout's rapid decline in the larger rivers. As brook trout declined, the state stocked millions of browns and rainbows to fill the brookies' niche. The introduced trout were wildly successful, even expanding into areas previously uninhabited by brookies. The reign of the brookie in the Catskills was over by the late nineteenth century.

After a quieter evening (I avoided bars), I headed out with a guide the next morning to fish the famed waters around Roscoe. I did not ask to fish for brookies, as I had caught plenty the day before. I wanted to follow in Theodore Gordon's footsteps, even if just for a day, and target some of the bigger fish in the bigger waters. Although I am a dedicated brookie angler, I am at times drawn to big waters and bigger fish.

I fished with Walter Ackerman, a well-known longtime local guide who lives on the banks of the Battenkill River and fishes out of Catskill Flies. Early in the day, we headed to Covered Bridge Pool, which had been a popular destination for Gordon, who, during the late nineteenth century, revolutionized fly-fishing in America by helping to popularize the dry fly, as opposed to the traditional wet fly, a method imported from England. His new technique, which he developed largely in the Catskills, created the practice of fishing for trout with flies that float on the surface instead of below it. Today Covered Bridge Pool still holds trout, but it has also become a popular picnic and recreational area, so fly-fishers should get there early on a warm summer day. We fished just above the famous covered bridge in a nice stretch of river with plenty of pocket water and submerged, suitcase-sized rocks. But I had no luck, not a strike, not a rise to anything I threw. It was early, and the previous night had been chilly, so we speculated the trout were holding tight. Besides, I wanted to fish in a few different areas, so we did not spend a great deal of time at the pool.

After an hour of casting to empty potholes, we set off for the Willowemoc, a tributary of the Beaverkill that is famous in its own right. Although I did not want to linger at Covered Bridge Pool, I could not help but look wistfully at the river as we drove away, because I would have liked to be able to claim I caught a fish at such a significant location. I don't know if I am alone in harboring this desire, but catching catch a fish in a specific location makes me feel a deeper connection with that landscape and its natural history.

Because this was one of many stops on my fishing journey north, I didn't realize at first that I would be in the Roscoe area during the Memorial Day weekend. I could not change any fishing dates at the last minute without messing up my entire schedule. I was amazed—naively so, I guess—at the number of people who appeared on the area's streams as the day wore on. Every roadside parking area, no matter how small, had cars in it. Larger lots had twenty cars, and one or two cars were squeezed into single spots. To avoid the crowds near parking areas, we walked down the Willowemoc to a quiet area out of sight of other anglers. The Willow, as it is sometimes called, has plenty of fishy-looking holding areas such as deep pools and slack water behind boulders, and it is big enough that nice size browns and rainbows ply its waters. This river and the Beaverkill must have been truly phenomenal when native brookies dominated 150 years ago.

Fly-fishers soon appeared above and below us. And boot prints in the sand and gravel told us that other trouters had visited these same fish earlier in the day. I had a couple of rises to my offerings but no fish all morning. I do not know if the crowds on the rivers were affecting the fishing, but I have to blame something. Even my guide seemed surprised by the lack of interest in my flies, commenting impatiently behind me as I cast, "Where are the fish?" It was a perfect day to be on the stream, though it was warm and sunny. Maybe too sunny? Maybe not enough wind? When fish are not biting, anglers mentally debate the reasons. Small groups of swifts chattered in the sky, reminding me that summer was here. And a glossy green tree swallow perched nearby, preening itself. Yes, it was a glorious day.

I can be an insecure and sometimes irrational angler. If I go a few hours without a fish, I tell myself I must be doing something wrong or my skills just stink. As a last resort, I convince myself that others have caught all the fish in a particular stream. I never acknowledge that maybe the fish just aren't biting because of any number of environmental influences, such as temperature, barometric pressure, or phases of the moon. We fished a few runs hard, meaning that I hit every trouty-looking spot multiple times with my fly. Finally, just before we broke for lunch, a stubby ten-inch stocked rainbow decided to be as hospitable as the local humans and ate my fly.

By early afternoon the Willow was really crowded. The parking areas were jam-packed, and bait fishers with over-sized bobbers were crowding a deep pool closest to the road. My guide just shook his head. Maybe I was right—all the fish had been caught. In the early afternoon, we headed to the Beaverkill. As at the Willow, many anglers were in all the well-known pools. After fishing for so long largely by myself, the crowds were distracting. Cars on the road along the river flew by with fly rods hanging out the windows and even poking from a sunroof or two. I had to remind myself it was a holiday weekend, and I was only two hours from New York City.

When I fish, I seek to be close to the landscape, not people. Fishing, hiking up a stream, and observing nature is at its heart a spiritual experience. It is really a selfish experience, too, in that I want to be alone. I do not want to share the experience or the water. Angling in the Catskills tested my ability to maintain my focus and even interest in my hobby/obsession. I never choose to fish in a crowded pool, an experience that had initially turned me away from the sport. That was why I felt no deep kinship with the Roscoe-area streams, no matter their history. They were too crowded to feel like home water.

High above the river Ackerman found a parking spot with barely enough room for a single vehicle. No path led to the river from there, so we made a slow and careful downhill stumble through the forest. My guide assured me he had never lost a client. I hoped the challenge would guarantee we would have the river to ourselves.

We waded across a wide shallow section of the Beaverkill to reach a deep swift run a few hundred yards downstream. No other anglers were in sight. Ackerman knew every fishable section of this river and, just as important, how to reach them. By the time we started fishing it was late afternoon; as the shadows lengthened, more and more mayflies and caddis began to appear on the river. This was a fairly treacherous run, as most of the river's water and energy fed into a chute, and the current was sweeping everything into that stretch, including me. My first instinct was to wade as close as I could to the edge of the deep run, because browns were feeding on the opposite bank. But if I got one step too close, I would have been pushed into the dark run. It was one of those challenging fly-fishing moments when I had opposite forces working against me: a strong current behind me that was trying to sweep me off my feet and trout in the distance. I stretched as far as I could in chest-deep water to reach them. All the while I somehow had to mend a great deal of line over a powerful current so my fly would float naturally.

Finally, after a solid long cast, I saw a distant splash on the opposite side of the run. I instinctively set the hook because, to be perfectly honest, I could not even see my fly.

Caddis fly (*Lepidostoma*). Drawing by Frederick Steinberg.

The resistance I felt told me I had a nice brown on my line. Several minutes later, Ackerman netted a meaty fifteen-inch brown. My trip to the Beaverkill was complete. It was not the most productive day I had ever had, but I felt lucky even to be able to fish such a storied stream, one that had been so abused but amazingly recovered through the initial work of those early fly-fishing conservationists.

The next morning at dawn, I set out under cool overcast skies for the Adirondacks. I was happy to be heading to an area with far fewer people and far more brook trout. I could relax again after leaving I-87 and heading into the mountains. I hoped the hand-to-hand combat fishing was behind me. The Adirondacks begin in the true north, in taiga-like country where boreal forest species begin to replace hardwoods, where long winters eliminate the faint of heart and leave those whose lives revolve around the opportunities nature offers, including fly-fishing. I had once thought I was one of those souls, but my academic ambition drove me out. I left New England years ago for academic jobs that offered smaller teaching loads and more research funds. Brook trout led me back to this region and made me question that decision, especially as I drove through a landscape with seemingly endless trout water. At times, I felt like this journey was my attempt to make up for lost time, to do in a summer what I could have been doing for years if I had not left New England.

I had passed through the Adirondack region many years before, on my way back to Missouri after spending the summer working as a house painter in Maine. I found the Adirondacks, like Vermont on that same drive, to be an ideal mix of wilderness, picturesque towns with fly shops, and endless all-season outdoor opportunities. I think these are ideal landscapes because you can fish for trout all day, return in the evening to a great dinner, and wash it down with a Snow Trout Stout or other local brew. I guess I value my creature comforts even if I like to fish in remote areas. But that is one of the benefits of focusing on brookies, which usually are found in areas that have also given rise to craft beers and opportunities to converse with outdoor-oriented folks.

The Adirondack Park is the largest park and state-level protected area in the Lower 48. It covers about 6.1 million acres, more than Yosemite, Yellowstone, Grand Canyon, Glacier, and Great Smoky Mountains National Parks combined. Adirondack has more than three thousand lakes and thirty thousand miles of streams and rivers, so fishing opportunities abound. The park isn't exclusively public;

instead, it is more like a biosphere reserve in that it mixes private and public land and has year-round residents within its boundaries. The state owns 2.6 million of its acres; the rest are in private hands.

Interest in establishing a protected area in this region began during the early 1880s, largely in response to the major environmental degradation occurring in New York State. As in the Catskills, the Adirondacks faced threats from industry, including widespread logging. Business interests at the time had a genuine fear that the Erie Canal and Hudson River might become unnavigable as a result of erosion caused by logging. The state formally established the park in 1892; in 1894 it designated all state-owned land within the park a forest preserve, which the state constitution declares "shall be forever kept as wild forest lands." Adirondack Park represents a unique public-private conservation effort, an idea that has been exported to other parts of the world and could serve as a model for the global biosphere reserve program so popular today.

For me, a key attraction of the Adirondacks is fishing its plentiful brook trout. Yes, the brookies had been chased to smaller streams, but because the park has thousands of miles of brookie streams, I doubted I would have any trouble finding them. To test my hypothesis, soon after entering the park, I spent an hour or so fishing a stream that runs parallel to the road. I do not even know the name of it, and it was not a recommended location; I just wanted to see if that trouty-looking stream any had brookies in it. Huge, elephant-gray boulders dominated the channel, creating innumerable small pockets of deep water on their downstream sides. Jumping from boulder to boulder was easier and more fun than trying to wade the stream because many of the pools were quite deep. I tied on a big fuzzy caddis and made a few casts. Pow! A brook trout. A few more casts farther upstream, and, pow, more brook trout flew out of the water after eating my fly. I knew then I was going to catch a lot of fish in the park. Apparently, my fishing karma was correctly aligned.

I was happy about my fishing success, but New York brookies have witnessed significant declines in 95 percent of the state's watersheds. The two most significant factors behind this decline are warming water temperatures and competition with nonnative fish species, including trout as well as smallmouth and largemouth bass. Both logging, which removes riparian shade trees, and an exploding beaver population that does essentially the same thing cause water to warm. A controlled trapping program and beavers' natural predators—coyotes, foxes, bobcats, otters, and great-horned owls—have not controlled the beaver population; they fail to reproduce only when the food supply is low, and they have abundant food sources in the park. As the beavers cut trees and build dams, the small reservoirs they create heat up and store energy more effectively than do streams. Often these reservoirs become too warm for brookies but are still tolerable for nonnative fish. Bass love them. Over time, the beaver population is likely to eat itself back into equilibrium. But by then bass and other warm-water species will have taken hold in local waters. New York is fortunate to have started from a point of great brookie richness. So while the decline is extensive, New York still has far more miles of brookie streams and ponds than most states.

I made the Hungry Trout Resort near Wilmington my base. Although the name was fitting for my quest, I chose the place because it sits on the banks of the West Branch of the Ausable River and offers brook trout fishing packages on privately owned water. Although the fishing press has not made the Ausable as well known as the Beaverkill, many anglers consider the Ausable a better overall trout stream than its southern New York counterpart. Some even claim the Ausable is the best trout river in all of New York. While designating a single river as *the* best seems a bit random, I liked the Ausable better than the Kill, because the former has more pocket water, hungrier trout, and fewer people. The West Branch is a big dark rough-and-tumble river that is a major challenge to wade when it's roaring in the spring and early summer. Like the Beaverkill, though, its days as a brookie stream are long gone, except in its headwaters. Today the Ausable is dominated primarily by brown trout.

I arrived early enough in the evening that I had an hour or so to fish. I walked down a short trail behind the hotel and reached the river. A few people were fishing, but the river and the area felt distinctly more wild than the Beaverkill. Large and small boulders strewn across the West Branch's channel had created thousands of small holding areas. I have no doubt that this river receives a great deal of fishing pressure from the tourists who flood the Adirondacks in the summer, but on that first evening, as the few other anglers packed up at dinnertime, I certainly felt like I was in real wilderness. Both March Brown mayflies and caddis flies were hatching. The hatch was not huge, but trout were rising around me. Small bats flying herky-jerky dive-bombed for flies over the river.

It was late dusk, not the easiest time to see fish, but I tied on my favorite March Brown imitation and gave it a shot. The only way I could see my fly at that point was to fish it nearby, with only a little line out. It was more like dappling than casting, but I didn't care; it was great to be out a river with bugs scattering throughout the approaching night sky. A few casts later in a nearby pool, I hooked a twelve-inch brown trout. Indeed, this part of the trip was humming along with good karma. I was doing something right. I cast for another twenty minutes or so but frankly was not sure if I missed fish or not. The sun was already behind the trees and mountains, so I called it a day and headed to dinner.

The next day, I fished with Jeff Kirschman, a longtime Adirondack guide who had fished the area since the early 1970s and worked out of the Hungry Trout. We headed to a group of interconnected ponds on private land about an hour away. This was not an inexpensive trip—getting access to private waters never is—but the opportunity to catch big brookies was just too tempting. How often do I get to the Adirondacks anyway? In such circumstances, "Bank account be damned!" is my battle cry.

The ponds resembled something you might see in Labrador. Dark spruce hugged the banks, while sparse, tundra-looking marsh bordered other areas. The sky was overcast, with strange layered stratocumulus clouds several thousand feet overhead. The formation resembled a huge vanilla seven-layer cake. On this cold gray day, I wore rain pants just for their warmth. I asked sarcastically, "What month is it?" Kirschman just laughed and told me I'd been in Alabama too long.

This was brookie fishing like no other I had experienced on this trip. Kirschman rowed me around the ponds, while I fished with sinking lines and streamers. I am by no means a dry-fly purist, but casting sinking lines all day to unseen fish is not the most exciting method of fly-fishing. While I find watching a trout clobber a fly on the surface is most exhilarating, I have caught more trout by fishing under the surface with nymphs and streamers. And this was one of those under-the-surface days, because it was midsummer and water temperatures were warming. Sensing my dashed dreams, Kirschman told me, "You can throw a dry all you want, but if you actually want to catch fish, grab the rig with the sinking line."

I truly appreciate guides who tell their clients the cold hard facts, even when we do not want to hear them. But what my day lacked in dry-fly excitement, it made up for in fish. It was a magical fifty-fish day, without a single one smaller than ten inches. Most were in the fifteen-inch range, and a few were even larger. I had gotten used to the six- to eight-inch fish on small mountain streams, so these brookies seemed like freaks of nature, or Frankenbrookies. I wondered if these ponds even had small fish. "Actually, I have a little bit of concern about that, because I never catch early-age groups in these ponds," Kirschman said. While the absence of a young fish could point to an ecosystem poised for problems in a few years, I was too giddy with success to put on my scientist hat.

The enormous brookies also stood out for their amazing array of colors. Many ranged from tanned leather to polished copper and were sprinkled with butter-colored spots, colors that matched the deep shadowy ponds. One was unlike any brookie I had ever caught: pale, almost steely blue-gray, with pinkish fins and large cream-colored spots without the usual halos. It did not look like a brookie at all, more like one of James Prosek's watercolors of a sea-run trout. While he took a break from rowing, Kirschman commented that not even he had ever seen a brookie like that.

Pond fishing for brookies was new to me, but it becomes more common as one moves farther north in New York and on into Maine, where such environments are commonplace as a result of glaciation. According to state biologists, New York State manages more than four hundred lakes and ponds as brook trout waters, and natural brook trout reproduction is known to occur in at least a quarter of these. The current number of wild self-sustaining brook trout ponds in New York State, however, is extremely low relative to historic conditions. A study published in 1990 reported that less than 5 percent of the 1,469 Adirondack lakes and ponds sampled contained unstocked truly wild brook trout populations. So while four hundred lakes and ponds sound like quite a few, they are a fraction of the number of wild brookie ponds that existed in the early nineteenth century when thousands held brookies. So even here, in a seemingly remote wilderness, human influence—in the form of acid rain and the introduction nonnative fish such as largemouth bass—is evident. Fortunately, the State of New York has established an active reclamation program for native trout ponds and has successfully reclaimed dozens. Again, it's a small number compared with what existed a couple centuries ago, but it's a valiant effort.

During my last day in the Adirondacks, I decided to fish a small tributary of the Ausable. Kirschman had pointed it out the previous day and suggested I fish it because I would have an opportunity to catch all three trout species. This accessible stream was open to both bait and fly anglers, so I knew that anglers probably worked the stream in droves. For the first few hours, I slowly made my way up the stream, hitting all the fishy-looking pools and undercut banks and casting around downed timber. It was fairly marginal fishing, even with the abundance of cover. A couple of small brookies were all I had to show for my hard work. The big fish of the previous day had raised my standards.

By late morning, I had covered enough ground that the stream had become no more than twelve feet wide. But it still contained a lot of small deep holes and more downed timber, so I wasn't giving up. As I came around a bend, I spotted another small deep pool, this one in front of a timber snag. The pool was no bigger than a hot tub, but the pile of timber and apparent depth of the pool told me it held a fish. I was less than ten feet to the left of the hole; because the stream was small and tight as a result of the tree cover, my cast was short and easy, a flip more than a cast. I expected perhaps an eight-inch brookie in that nice pool, so when a twenty-inch brown trout rose from the black water and ate my fly, I almost fell over. It actually scared me as it rose from the bottom of that black hole and lunged at my fly, its black, almost shark-like, eye rolling at the water's surface. I would never ever have suspected such a big fish in that small stream, let alone one that would appear out of nowhere to take my fly. I hadn't had much luck all morning, so I was fishing pretty sloppily at that point—stumbling, making bad casts—which contributed to my amazement.

Because the stream was small, the fish did not have much room to run. And this was fortunate, as I was fishing a light three-weight brookie outfit. The fish shot back and forth in its small pool, while I stood there hoping it would not dive under the snag and break off. It did not strip much line out, since it really had no place to go, with shallow riffles above and below its small sanctuary. The fight lasted only a few minutes, and I was able to steer it to a gravel bar—or, just maybe, did the fish swim up onto the bar? It's all a blur. It was a handsome maize-colored brown trout with ruby-red dots on its sides.

As I removed the tiny hook and released the fish, I felt sort of dazed and confused. The fish had appeared so quickly, then I had caught and released it so quickly, that I asked myself—aloud, I believe—"Did I really just catch that fish in *this* stream?" While that catch was exciting and I can celebrate the memory, finding a big brown like that in such a small stream provides a window on the challenges brookies face in waters they share with introduced trout. I have little doubt that that big brown grew to that size by eating most of the brookies in that stream, and that trade-off is hard to celebrate.

PART III

New England and Labrador

CHAPTER 11

Connecticut

I WAS SITTING in my car on Interstate 84 near Hartford on a Friday evening in August. Traffic was literally crawling. People were escaping work, people were heading out on vacation, and everyone seemed to be leaving Hartford, Waterbury, and any other urban center in the state. And all seemed to be competing for space on whatever stretch of highway I found myself. My hands were locked to the steering wheel, sweating, because I gripped it far too tightly. I was uptight, I was wasting time. I felt thousands of miles from brook trout. I was crisscrossing New England, heading to Newtown—yes, site of the horrific 2012 shooting at Sandy Hook Elementary School—in the southwestern part of the state to meet with Joe Hovious, who was then the conservation chair for the Candlewood Valley Trout Unlimited Chapter. (Today he serves as vice president of the Pootatuck Watershed Association.) Leaving the solitary world of trout streams and entering the humanized concrete world was almost overwhelming. Although Newtown is a quaint small town that was founded in 1705, it is now part of the larger urban Northeast where people dominate the landscape through subdivisions, malls, cars, and roads.

Much of Connecticut is emblematic of the difficulty of maintaining brookie populations in and around urbanized landscapes. Anyone who has driven on Interstates 95, 84, or 91 probably has experienced the snarled traffic jams creeping through the impervious urban landscapes that are so common in not just Connecticut but in much of the Northeast. And along with this tide of humanity come our demands on water. We draw huge quantities of water into our urban areas like giant sponges and always release it back to the natural world in poorer shape and smaller quantities.

Just when my grip on the steering wheel couldn't get any tighter, I saw the exit sign for Newtown. As I left the highway and headed toward the town's center, I felt a sense of relief, for colonial architecture and ancient maple trees have greeted me. As I slowly drove the quiet streets of the prosperous town of neatly kept lawns and flower gardens, I was able to breathe again, just as the shadows grew longer and the cicadas began their familiar and comforting summer evening chorus.

I chose to visit part of Connecticut, because it represents the front line between urban development and cold-water conservation—an urban-environmental eco-tone. Certainly, this is not the only location, or even state, where such battles take place. Eastern Maryland, for example, is rife with similar landscapes, but in this part of Connecticut, trying to maintain a balance between concrete and green spaces is especially fraught. This specific landscape also attracted me because of the dedication of the local Trout Unlimited (TU) chapter to its cold-water resources. I met with plenty of motivated TU chapters during my travels, and the Candlewood Chapter was second to none. Its activities ranged from Trout in the Classroom projects to riparian restoration efforts. All involved the local community and, especially, school kids such as those at Sandy Hook Elementary.

I met Hovious for breakfast at a Newtown diner before we headed out to visit a couple of nearby streams. "The impacts of urban sprawl have really devastated small streams in this part of the state," Hovious told me. "We have lost a lot of our green spaces and riparian forests, so our trout streams really struggle during the late summer because of the higher water temperatures. Our chapter has worked on streamside tree-planting projects, as well as trying to preserve as much green space as we can in the area. And it's not just TU; there is a lot of local interest in keeping remaining green space from being developed." This teamwork approach seems especially critical in areas such as Fairfield County, where real estate prices are especially high, making the purchase of important properties difficult for local conservation groups.

After breakfast and some fish stories, we headed out for a tour of streams on which the local TU chapter had been focusing its protection and restoration efforts. We were in the car for only a few minutes before we reached Deep Brook. Its proximity to Newtown and the interstate was the first tip-off to the challenges that local conservationists face. Deep Brook suffered home heating oil spills in 2003 and 2004. Both originated from underground fuel tanks, and both resulted in fines by the Environmental Protection Agency (EPA). Trout Unlimited was participating in the restoration program, monitoring for the presence and diversity of aquatic insects and improving riparian cover through tree-planting projects.

Underground storage tanks do corrode and sometimes leak, so in parts of the East where these tanks have been in the ground for decades, spills are inevitable. And local streams' proximity to the tanks made them especially vulnerable. The oil has long since been cleaned up, but the stream continues to suffer from the initial spill, with fewer fish and aquatic insects. If such an event can have a silver lining, it is that the town, with the help of the Candlewood TU Chapter, has become more aware of water quality and conservation issues. The spills focused the community on the vulnerability of streams in urbanized landscapes and the need for better protection. Another benefit—and I use that word loosely—is that the EPA gave a portion of the fines to the local TU chapter for monitoring and restoration. "We are able to better understand the long-term impacts of the spills," Hovious said, "rather than just wondering what might or might not be happening. And we can get the community involved through these projects, which hopefully creates a [greater] appreciation for our local streams and water supply. This is especially important with the school groups."

Our next stop was the Pootatuck River, a larger stream whose watershed includes Deep Brook. Because the river is larger, its vulnerability to pollution is greater. As it did with Deep Brook, the Candlewood TU Chapter has made the Pootatuck a focal stream, initiating bank stabilization and streamside tree-planting projects, as well as an ecological monitoring program. The Pootatuck, or at least part of it, was a classic urban stream, even running beneath I-84. Although the section of the Pootatuck that I toured seemed like anything but a trout stream, with big rigs screaming along the overpass above our heads, the stream was clean. I saw no tires, shopping carts, or plastic bags clinging to shrubs. TU's volunteers and others had done a great job of holding back the urban monster. "Development is such a big challenge here," Hovious said. "Look around. We really have to fight to keep every piece of open land undeveloped."

I visited the area in early August, when water temperatures were marginal for trout. A new parking lot or strip mall that delivers hot runoff in the summer could tip the balance for many streams. In densely settled areas in Connecticut and the Northeast in general, the buffer between pavement and fish is often little or nothing. As we drove around the Newtown area and its outskirts, Hovious pointed out a few recent developments that had contributed hot stormwater to local streams. One parking lot was especially bothersome, because it literally abutted the river. As we inspected it, Hovious said, "Just think about how hot water can become on this parking lot at this time of the year. And all that water makes its way into the river."

After we parted ways back at my car, I returned to both streams to fish for a few hours. The sun was high and the water warm in the late afternoon, so I did not expect much. Sometimes expectations are met. I raised no fish on Deep Brook or the Pootatuck, but I do not blame the oil spill or hot-water runoff. Water that is hot and low can mean really slow fishing, and I fished fast, racing against the evening shadows. The air was humid and thick with dragonflies darting above the stream. I felt that I had not given the streams enough time, but at least I was able to tour the area and see the efforts under way to make sure trout will still be in those streams the next time I visit.

The next morning, I loaded my car at dawn for my trip to Salisbury in the northwestern corner of the state. A mockingbird sang its dawn song from a nearby shrub. It seemed late in the summer for the bird to be looking for a mate. I welcomed the company. The town slept and the air was still cool as I drove through its empty streets. I thought about brook trout around Newtown. I wondered whether those streams would have brookies in them in fifty years. The development pressures won't subside anytime soon. These pressures are what global environmental change looks like on the ground, up close.

It is easy to feel overwhelmed and become paralyzed by larger-scale environmental problems. "What can one person do?" is a frequent refrain. Well, in Newtown, one person planted a tree, removed junk from streams, and shared their enthusiasm for trout with a school kid. The more I thought about it as I drove away, the more I actually liked the brook trout's chances.

The practical results of my poor fishing around Newtown meant I was still without a brook trout in Connecticut. I was headed to Salisbury to meet with Kirt Mayland, who was Trout Unlimited's director of eastern water projects. Northwestern Connecticut is a totally different landscape from that found in the eastern half of the state. I passed mature woodlots and horse pastures, old farms and rolling green hills. There was no need for white-knuckle driving. I reentered my comfort zone.

I met Mayland in a small coffee shop at the back of a pharmacy in downtown Salisbury. At the time, he was focused on monitoring water policies in the Northeast and was especially concerned about the drawdown of streams by towns and industry. He told me the streams had little protection; towns and industry could more or less drain them when they needed water. The water laws on the books were designed and enacted during middle of the nineteenth century, before the expansion of suburbs, ski slopes, and golf courses. So even though demand for water has increased, laws have not changed to safeguard the environment. As the summer heats up, the reduction in stream flow can and does have serious consequences for temperature-sensitive species such as trout.

I was aware of the water battles in the parched West but was surprised by the extent to which similar issues exist in the Northeast. I had naively thought this part of the country had plenty of water for everyone, especially in the green countryside. Not so, according to Mayland. For a report titled *A Glass Half Full* (2006), he and others documented the increasing and varied consumers of water in the Northeast. Demand for water obviously increased as populations grew, but residential users in ever-expanding suburbs are not the only ones who drain rivers; according to the report, ski resorts use significant amounts of water to make snow, especially in Vermont, New Hampshire, and Maine; bottled water companies withdraw from aquifers that also feed trout streams; and golf courses and farms use copious amounts of water during the summer when natural flows are already at their low point.

As we talked, I still could not fathom that rivers literally run dry in New England. According to TU, in 2004 Connecticut identified more than sixty rivers and streams that suffered from "flow impairment." Similarly, Massachusetts had 160 rivers, Vermont had 50 rivers, and Rhode Island 35 rivers that flow reduction had altered. In one especially distressing example from 2005, the Fenton River in Connecticut was sucked dry when college students returning to the University of Connecticut in Storrs turned on their faucets and showers at the beginning of September. "Around eight thousand brown trout were killed," Mayland said. "The river literally dried up when the students returned. It was the sort of large-scale event that really crystallized this problem in Connecticut." In the years since the TU study was published, demand and pressure on rivers have only increased.

Today Connecticut and surrounding states are faced not only with warmer streams, as paved surfaces continue to spread with the advancing suburbs, but also less water, as many of these same developments draw from streams. Trout Unlimited has increasingly focused its political clout on changing laws at the state level to guarantee minimum flows in the region's rivers and thereby safeguard aquatic life. It seemed especially ironic that, as we had this conversation and drank coffee, a torrential thunderstorm was raging. On that day, we had more water than we wanted.

After a couple of coffees, Mayland took me to a nearby brook trout stream to show me what he was protecting in Connecticut. I had heard so much about streams with little or no water that I needed to fish. Bad environmental news always drives me to seek respite in streams and brook trout. And, storm or no storm, I still needed to catch a brook trout in Connecticut. Mayland had another meeting to attend, so he promised to point me in the right direction.

With directions in hand, I drove a short distance to a nearby trailhead, then hiked up a steep gorge. Huge boulders, exposed rock shelves, and cliff faces dominated the stunning landscape. Between the weather and the old hemlock canopy, it was a dark walk, as if the gorge and its slick walls were protecting those deep brookie pools. The ground was saturated with rain and humidity, making the climb down the stream treacherous, and huge drops of cold rain fell from the canopy, splattering my head and neck and sending chills down my spine. Rappelling equipment would have been useful for reaching the most remote pools. This stream, the precise location of which I will not divulge, had never been stocked, according to Mayland, so it represented the best possible angling for native fish whose genes have not been tainted by state hatcheries. The smell of wet earth, moss, and rotting vegetation floated thick in the air. It was a good healthy smell.

Even with the heavy rain that morning, the stream ran gin clear, the sign of a forested watershed and good riparian cover. It consisted of one deep pool after another

that spilled down the gorge, an absolutely perfect brookie stream, large enough to provide casting room, deep pools, and hungry fish. I started with an old standard, the woolly bugger. After just a few casts to the first deep pool, I brought in a pretty eight-inch fish. A few more casts, and I landed another fish. *Now*, I thought, *this might be a fifty-fish day.* I soon switched from streamer to dry fly to mix it up a bit. While the fish were not as eager to leave the safety of their deep pools to take my dry fly, I still caught plenty of trout, as I moved from pool to pool. I had my Connecticut brookie and then some.

After fishing a few more hours, thunder and lightning began to crack violently around me. Because of the forest canopy, I did not see the storm coming. I have fished through many rainstorms, some cold driving rain, but this was the sort of summer thunderstorm that made me feel the reverberations of nearby lightning strikes deep in my body. The air smelled of ozone, as what little sky I could see through the dark canopy turned black. I accept karma, so perhaps my departing this world on a trout stream would be fitting, but I was not going to stupidly keep fishing. The skies opened, and I ran to my car, keeping my rod tip as close to the ground as possible to avoid being struck by lightning. As I sat in my car, mothball-sized hail began to fall. It was time to move on. I had caught plenty of gorgeous little Connecticut brookies; I had no need to push my luck.

As I look back on the streams I have visited and fished in my quest, I have often asked myself whether I would want to return to fish a specific location. That stream near Salisbury is one of the several spots I would happily revisit. Many things figure into our attraction to a landscape. Certainly for me the fish play a role, as do scenery and crowds (that is, the lack thereof). This particular stream was not remote. I did not have to work to get there, so it was not the challenge that attracts me. I parked on the side of the road and caught trout a few yards away. But that is one of the reasons I found it intriguing. The stream is located near a busy state highway, but it is full of brook trout. Local certainly people knew about it, yet I saw no beer cans or cigarette butts or worm containers, just brook trout. I wish all trout streams were treated as kindly.

The trail along that stream eventually provides access to the Appalachian Trail, the same trail I have hiked in various states during the past twenty years and where I fished and camped in northern Georgia a thousand miles to the south. My geography and that of the brook trout were once again deeply interwoven. Their geography remained my geography.

CHAPTER 12

Massachusetts

AS I WAITED in front of my hotel in Boston's Back Bay neighborhood, it felt odd that I was about to go look for brook trout. I was in the city for an academic conference in 2008, not on my usual journey north from Tuscaloosa or south from Maine. But there I was, getting into a car with Michael Hopper, a brook trout activist, and my waders and angling gear. The bellman did not know what to think of my pile of gear. He kept asking me if I needed help or directions. We quickly passed through various back streets, trying to avoid congestion in the heart of the city, and got on Interstate 93, where we passed various apartment complexes that were built literally next to the shoulder of the highway. Finally, we broke free from traffic and the concrete as we headed south.

I knew western Massachusetts had brook trout, but as signs for coastal locations such as Hyannis Port and Cape Cod National Seashore whizzed by, I had beaches, sailing, and the Kennedys on my mind, not brookies. I voiced these thoughts. But Hopper assured me that many of the streams we were passing over on the interstate once held brook

trout. "Daniel Webster fished these streams," Hopper said, adding that Cape Cod "was a destination point for Boston fishermen in the nineteenth century. Before the mills, before the farms, these streams were home to big brookies."

Even the physical landscape looked anything but trouty. As we entered Cape Cod proper, the landscape was dominated by scrubby pine forests. The terrain looked almost xerophytic (like a desert), with beige-colored dry sandy soils, not the lush green forests I normally see in New England. The spruce, hemlock, and maples are farther inland and on higher ground with more soil. Those are the trees I associate with brook trout. Instead of forest birds, I saw seagulls and ospreys.

Cape Cod and Long Island represent the southern terminus of a giant glacial moraine created by the last ice age. Whereas Long Island's soils are generally good for agriculture, the Cape received huge deposits of sand. Thus, the dry scrubby look is no accident, just the result of the porous sandy soil. My internal trout compass felt like I should be heading west toward the Berkshires and not south to the beach. But therein lies the uniqueness of Cape Cod's brookies: They are special because they are sea-run fish, earning their nickname, "salter" brook trout.

Hopper's enthusiasm for the salter and various restoration efforts on its behalf was obvious as we drove. Like me, he is a brookie fanatic, and he has directed his energy to helping to create the Sea-Run Brook Trout Coalition, whose purpose is to restore and protect sea-run brookies on Cape Cod and in other coastal rivers of the Northeast. Like so many rivers and streams in the region, pollution, the construction of dams, and channelization have affected salter brook trout streams. The Industrial Revolution in the Northeast was not kind to trout streams and rivers, especially coastal ones, where so many people settled.

We had traveled to the Cape to fish the Quashnet River for sea-run brookies and to check out a long-running restoration program. The Quashnet suffered from the effects of industrial development after it was dammed for a mill that polluted its waters during the nineteenth century. Then, beginning in the 1890s, the river was significantly degraded even further by the conversion of the landscape to large commercial cranberry bogs, which had a dramatic effect on the river's hydrology and its watershed. For example, by 1900, the Quashnet's entire valley—from Waquoit Bay, located on the lower "elbow" section of the Cape, to its headwaters—was stripped of all its vegetation for hundreds of feet along the river, which increased water temperatures and silting. Also, the surrounding dunes were leveled to make it easier to create the bogs. Channels were dug to control water around the bogs, and dikes were constructed to help flood them. Annually, tons of sand were spread on the bogs to control the unwanted growth of vegetation, and some of the sand made its way into the stream, inundating the remaining spawning beds for brook trout. Even the main channel of the Quashnet was straightened, increasing the scouring force of the stream, which reduced in-stream structure such as sunken trees, snags, or other debris.

By the mid-1950s, the coastal stream once thriving with brook trout, herring, and eels was all but dead. In roughly half a century, cranberry interests essentially took a meandering spring-fed trout stream and turned it into a pond no longer fit for brook trout—or most fish species, insects, and plants, for that matter. In all my travels, I don't think I had seen or heard about a more altered stream, and I doubt I can ever again consider cranberries a healthy snack, given their environmental toll. In 1954, however, the Quashnet received a reprieve when two hurricanes damaged the cranberry bogs and led to their closure.

With the sudden demise of the cranberry bogs, the slow rebirth of the Quashnet began. The state purchased a narrow riparian corridor along the river in 1956 and, during the 1960s and 1970s, began stocking the river with nonnative trout. Surveys found that the stocked fish were not reproducing, so the state abandoned this program in 1991, thereby setting the stage for the eventual return of wild and native brook trout. Although cranberry farming had largely destroyed the stream's ecosystem, the state's fisheries officials targeted it for restoration during the 1970s. That decision prompted volunteers from the Cape Cod Chapter of Trout Unlimited (TU)—under the direction of Fran Smith—to spend thousands of hours cutting back the sweet gale that then dominated the riparian forest, rebuilding the stream banks, and restoring the natural channel and meandering bends. Some of this work continues today.

One of the more interesting components of TU's restoration work is building overhanging banks that resemble shelves. These are constructed with boards that eventually support natural vegetation on their surface, creating an almost perfect hiding spot for trout. Hopper called them "trout condos." The volunteers' commitment to restoring the stream to a more natural state also helped fend off development pressures during the 1980s and eventually led to the creation of the Mashpee National Wildlife Refuge in 1995. With the physical restoration of the stream largely complete, the brook trout followed. Genetic tests show that the salter brookies present in the Quashnet today are genetically wild fish, not descendants of stocked fish, so the early stocking efforts had no lasting effect. The latest research effort, partly funded by the local TU chapter, is to track the long-term movement of fish in the river, using passive integrated transponder tags. Joined by other local conservation groups, these TU members also have also worked to restore other salter streams, such as the nearby Red Brook, but I am highlighting the history of the Quashnet because trout conservationists, especially Fran Smith and a handful of local trout enthusiasts, began to restore salter populations and habitat during the 1970s, long before the state developed or even acknowledged plans to conserve brook trout populations in other areas. They were truly ahead of their time. And until one visits the river, it is difficult, if not impossible, to understand how much time and labor they spent on this project.

While brookies inhabit only a few coastal rivers and streams today (nine in all of Massachusetts), they survive because of the most organized and intensive trout restoration efforts in the eastern United States. Providing the history of this or any restoration effort in a few pages is easy, but describing the emotional, financial, and physical commitment such long-term restoration efforts take is impossible. One reflection of this commitment is that the National Oceanic and Atmospheric Administration awarded Fran Smith its Environmental Hero Award in 2003. This honor recognizes volunteers for their "tireless efforts to preserve and protect our nation's environment."

As Hopper and I drove back to Boston, he both lamented the loss of so many salter streams and expressed excitement about the prospects of future restoration efforts. "If you read [Jerome Van Crowninshield] Smith's 1833 book, *Natural History of the Fishes of Massachusetts*, it is just amazing how many of these streams had salters in the early 1800s. It's hard to look around today at so many of these streams without becoming somewhat depressed. But, on the other hand, look how Red Brook and the Quashnet have come back. That is something to be excited about and shows what hard sustained work can accomplish."

Fast-forward more than a year, and I am on my second trip to Cape Cod, this time in early August on my way south to Tuscaloosa. I did not actually catch a brook trout on my first trip with Hopper, so I was unable to check Massachusetts off my list. This time I also met with Mark Hattman, who, at the time, was president of the Cape Cod TU Chapter. It is striking that even protected trout streams flow so close to development in the Northeast. In the case of the Quashnet, we parked and then hiked in a short way from a nearby condominium development. In the Northeast, development pressures constantly threaten or ultimately eat away at the edges of green space. This pressure is especially intense in such highly desirable areas as Cape Cod, making the rebirth and protection of the Quashnet all the more impressive. Hattman agreed with my perception, saying, "As you can imagine, this is valuable real estate. Development pressure just doesn't go away, given its value. So conservationists face some serious challenges around here."

Fly-fishing on these small salter streams presents a different set of challenges than on the typical brookie stream in the mountains, which requires hiking up a stream and casting to one pool after another. These streams instead require downstream fishing in most locations. Like fishing the tunnels of rhododendron in the southern Appalachian Mountains, angling in the Quashnet does not lend itself to graceful long casts because of the tight quarters. Short rods and light leaders are the requisite tools here. You have to allow your fly line and streamer to drift with the current into a deep pool or undercut bank, whereupon you then slowly strip the line upstream. This approach is not that different from fishing deep pools on large rivers, where the fisher allows a streamer to sweep downstream, but, considering the size of the Quashnet, at first it seemed strange. Small streams are meant to be hiked and fished against the current, or so I thought. The streamside shrubs and trees were also so low and tight that actually casting a fly would have been difficult, if not impossible, in most sections of the river.

As we fished, Hattman reiterated the significance of the restoration work: "In the thirty-odd years this has been going on, it is one of the oldest restoration projects in TU history. The true significance is that the success of the restoration has allowed a vibrant population of sea-run brook trout to be here today for the scientific community to work with. Without the work done decades ago, there would be no native nonhatchery fish to study."

As I stripped my line, it struck me how separate and isolated the river felt from the bustle of summertime Cape Cod just a short distance away, where traffic jams dominate the roadways as summer vacationers crawl their way to the beaches. The vegetation, coupled with the cold water, creates a microenvironment that feels, looks, and smells different, something like cool moss or soil after a thunderstorm. Yes, we were near strip malls and condominiums, but the streamside vegetation and cool air that floats above the water was a welcome respite from civilization. The water was amazingly cold for early August, reminding me why a few brookies were able to hang on and repopulate, even as the stream was largely destroyed by cranberry farming.

Like so many other folks with whom I have fished during my travels, Hattman insisted that I fish first and take the best stretches. I did not protest, because I needed to catch a brook trout for my list. I could have easily headed to western Massachusetts to satisfy my quest, but I chose to return here because of the history of this stream and because salter brookies are unique. As I let my line drift toward a promising pool, Hattman acted as guide: "OK, strip, faster, OK, get ready . . ."—then, bam, a fish struggled against my line and the current. A few minutes later, an eight-inch brook trout rested in the palm of my hand. Its color was a darker, more coppery orange than I expected.

When I think of a sea run, I think of fish that are almost washed out. But this fish was small enough that it probably had not yet ventured to the sea, so I should not have been so surprised.

As we continued to fish downriver, each of us caught several small brookies. Each fishy-looking spot usually produced a brookie or at least a strike that was exponentially more violent than the actual size of the fish. No, these were not the twelve-inch-plus fish that Hattman and others have found and caught during stream surveys, but that the Quashnet held any brookies at all was mind-boggling and made the trip worthwhile.

As we left the stream in the early afternoon, I felt a deep sense of gratitude. Catching brook trout in a stream where few or none existed in the recent past is a remarkable feeling. I experienced the same euphoria in Pennsylvania, West Virginia, and Tennessee, where I fished steams once deemed dead or severely degraded. I was grateful for the resilience of nature and the restoration pioneers whose unwavering commitment continues to pay off.

CHAPTER 13

Vermont and New Hampshire

ROUND BALES OF yellow straw, set against a green mountain background and black-and-white Holsteins, shared this part of the White River Valley with me. Clouds of flies and even a few dragonflies buzzed the cows, who stared nervously at me, their shiny hides twitching in the sun. Flies launched into the air each time they moved. I was glad the cows kept their distance; I wanted no part of that insect frenzy. As I sat by the bank of the White River and leaned against a maple tree on the edge of the pasture, small rainbow trout rose in a nearby pool, sipping tiny midges from the river's surface. Swallows, which probably were nesting in a weathered barn downstream, performed acrobatics over the river, competing with the trout for their evening meal. The air was ripe with the earthy smells of summer—animals, sweet-smelling mowed grass, and the mossy, heavy aroma of the river. I did not disturb the trout; I was content to breathe in the beauty around me.

Returning to Vermont is a homecoming of sorts, because it is where my love for brook trout and their landscape began. I return here to fish as often as possible, so I know some of its streams better than those in any other eastern state, and I know best those in the central part of Vermont, where I have spent many days hiking and fishing in the Green Mountain National Forest. Thus, Vermont always feels familiar. To be sure, Vermont's fish are not the biggest; most are quite small. So, while in Vermont, I am never after trophies, just the love of fishing in these home waters.

In addition to the countless brook trout streams, I am drawn to Vermont because of its cultural landscape. Couple its small farms, dairies, artists, writers, craftspeople, and a culture that embraces trout streams, organic cheese, sausage, and local beer, and you have my proverbial home away from home. A drive through most of Vermont is also a drive back in time, before the era of big-box stores and fast-food chains. Instead of Walmarts, and CVS stores, and McDonald's, country stores, which sell everything from sandwiches to dry flies, remain a fixture of rural retail. If country stores do not have an item, it's probably not critical to daily life.

However, Vermont is a tough place in which to make a living, with its long winters, difficult job market, and small population. Many folks who call it home truly want to be there, and they have to be creative to make life work. Many, especially in rural areas, have to work a variety of jobs. This has resulted in high levels of poverty, about 10.2 percent in 2020, according to the U.S. Census Bureau.

Many years ago, when my children were young, my family and I pulled into a Vermont rest stop on Interstate 91 and found a wide assortment of locally made products on display. The visitor center offered everything from handmade wooden canoes to organic goat cheese. Outside the building, local community groups were holding a huge bake sale. When we ran out of small bills, elderly women in plain rag-wool sweaters gave our kids free cookies and brownies. Were we really at an interstate rest stop? It looked more like a country fair. My kids loved it and gobbled down as many chocolate goodies as they could, while I tried to drag them back to the car to avoid the inevitable sugar high and subsequent crash. They still talk about that rest stop with free treats. Vermont maintains an increasingly rare sense of place, different even from other New England states'.

Vermont was also home to Robert Frost, among America's most recognized poets. Although Frost spent most of his life in New England's brookie country and wrote extensively about the human role in nature, he apparently was not an angler—or, at least, not an avid one—because only one of his poems, "The Mountain" (1914), mentions trout fishing:

> You've never climbed it?
> I've been on the sides
> Deer-hunting and trout-fishing. There's a brook
> That starts up on it somewhere, I've heard say
> Right on the top, tip-top, a curious thing.

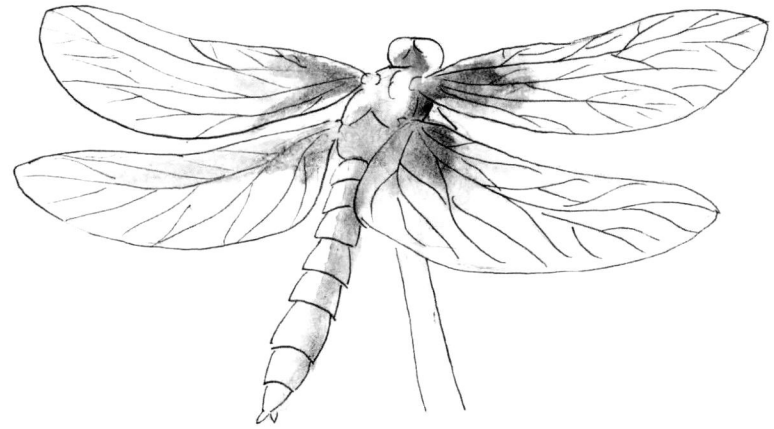

Dragonfly (*Odonata-Zygoptera*). Drawing by Frederick Steinberg.

It almost pains me to think that brook trout played no role in Frost's writing life. How could he not be drawn to brook trout, given the many streams around his New England home? Nonetheless, he remains my favorite poet, because he captures in words the soul of the New England countryside. I feel as if I know the landscape even better after reading his work. Perhaps in order to know Frost better, I fished for brookies in a stream along the Robert Frost Commemorative Trail in central Vermont. Given its lack of in-stream cover, the stream is fairly marginal for brook trout, but the few small fish I caught were worth it.

As I walked along the trail that led to the stream, various poems displayed on placards supplied not only inspiration for the day ahead but also a means to think about the landscape around me. His poems appeal for both their rural New England setting and their emphasis on the human struggle with nature. Reading them aloud in the presence of crumbling stone walls and abandoned fields, I could almost feel Frost's presence. He had the pulse of rural New England, even if he did not share my obsession with brook trout.

Like the Catskills of New York, Vermont represents a center for fly-fishing in the eastern United States. One July, I found myself in Manchester on a historical and retail pilgrimage. The elegant small town is filled with stately inns and private homes, all painted a crisp white. But beyond the eighteenth- and nineteenth-century colonial architecture, I was there to visit Orvis. The number of fly-fishing equipment companies has exploded in since 2000, but Orvis, founded in Manchester in 1856, remains a giant in the field. It sells far more dog beds than bamboo rods these days, so it is not exclusively a fly-fishing company, but it remains iconic. I went to its flagship store in Sunderland to look around, test a fly rod or two, and buy a few flies. Instead of a classic fly shop, though, I found an outdoor-oriented retail cornucopia. I somewhat expected this, based on the Orvis catalog, which is filled with everything from luggage to sport coats, but the scale was a little overwhelming, with hundreds of tourists milling about the large bright showrooms and trying on one pair of gaudy pastel pants after another.

The store did have a nice selection of flies, which probably do not sell quickly. I doubt many locals buy flies here. During the summer and fall, when the place is packed with busloads of tourists, most locals probably avoid the congested shopping areas that dominate the area. I don't begrudge Orvis the sale of dog beds. It has created a niche that caters to a high-end sporting lifestyle and reaches an audience far beyond fly anglers, which is why Orvis has survived while other fly-fishing equipment businesses come and go like the seasons. The company also generously contributes to conservation causes, so it is committed to the larger fly-fishing world and the resources that sustain it.

Many years ago, I even tried to get a job with Orvis. As I approached graduation from the University of Missouri in 1988, I contacted Orvis about a retail job selling fly equipment at its Manchester store. The fishing manager was friendly but told me the store had no openings and, in fact, had a waiting list of many people for a position. I guess I was not the only angler who thought living in Vermont sounded good. That could have been a good life. Even with the crowds, being in a real center of fly-fishing history was great. The cabinet of handcrafted blond bamboo fly rods made me salivate. Given that each costs more than $2,000, I wondered how often someone comes in and spontaneously purchases one.

Across the parking lot from Orvis is the American Museum of Fly Fishing. Located in a converted and restored Victorian home, its displays celebrate artifacts from the history of fly-fishing and technological innovations. Antique tackle, ancient bamboo rods, and fine artwork depicting grand days on streams instilled in me a wistfulness for a pristine nineteenth-century stream. The museum also publishes a first-rate journal, the *American Fly Fisher*, that runs many fascinating, albeit obscure, historical fly-fishing accounts. Both the museum and Orvis provide Manchester with historical anchors in the fly-fishing world.

Vermont also ranks high on my list of states that have contributed to the fly-fishing literature. W. D. Wetherell and Craig Nova have penned works about or related to fly-fishing in the Green Mountain state. While bookstores probably shelve their books in the nature section, both are far from typical how-to books. They use fly-fishing as a backdrop for examining larger issues in life. Wetherell's collection of essays ranges from exploring home waters throughout the seasons to a hilarious description of his brief career at a large unnamed tackle company (Orvis?). In that essay, Wetherell insists on demonstrating his casting prowess to a blowhard costumer, whereupon he wraps the fly line around a bench, shattering the rod in his hands, as his back cast meets inert resistance. It is probably the funniest fly-fishing essay I have ever read, because what fly-fishing addict has not dreamed of working at Orvis?

Craig Nova, on the other hand, uses fishing for brook trout as both a retreat and a source of inspiration through such critical phases of life as dating, marriage, raising children, and coping with financial challenges. When I assigned his book in my class on outdoor literature, my University of Alabama students often would ask me why, because it's not a fishing book. Perhaps it is not so apparently relevant to a nineteen-year-old, but for a fiftysomething father who seeks solace in nature, it is the best kind of fishing story.

No trip to Manchester is complete without fishing the nearby Battenkill. It has earned the reputation of being home to some impressive browns, but these fish are decreasing in number and are increasingly selective regarding flies. The Battenkill is a melodious river that meanders through the Manchester area and surrounding bucolic countryside. A tour of this river really is a tour of Vermont in that it courses through a slice of every piece the state's land and life: Manchester (small town), fields with dairy cattle and corn (agriculture), covered bridges (history), and Norman Rockwell's home in Arlington (art) before entering New York and eventually joining the Hudson River in Schuylerville.

This patchwork landscape, however, means multiple threats to the river. Studies have documented a clear decline in the number of trout in the Battenkill. The single most significant factor in this decline is the lack of instream woody debris. According to the *Batten Kill Trout Management Plan, 2007–2012*, agencies have conducted numerous studies to identify likely causes for the reduced abundance of wild browns in the Battenkill; results from these investigations point to habitat deficiencies that have been in the making for many years. In response, groups such as the Battenkill Watershed Alliance, state fisheries personnel, and officials with the Green Mountain National Forest have worked together to improve fish habitat in the river by placing large trees in appropriate locations and to educate local landowners about bank stabilization and riparian cover issues. While it is perhaps premature to declare total victory, fish surveys since 2012 have found more young trout, indicating a positive response to habitat improvement projects.

A multitude of other factors also have had a deleterious effect on trout populations in the Battenkill, including increased urbanization around Manchester, agricultural pollution, and such ecological changes as the population explosion of trout-eating mergansers. The day I fished the Battenkill, I saw a female merganser with eighteen chicks. I wondered how many trout that one bird ate to raise her chicks. Trout and mergansers have coexisted for a millennium, but when other factors stress or reduce trout populations, they magnify the effect of hungry mergansers. Changes in fish populations and fishing conditions also are the result of the large numbers of inner tubes and canoes that converge on the river and disrupt fish behavior. I imagine that many of those floaters move downed tree limbs and larger timber out of their way and unknowingly step on and destroy spawning beds. I had to leave the stream to avoid multiple flotillas of brightly colored rafts dragging beer-filled coolers as they made their way slowly down deep trouty-looking runs.

I had no luck during the afternoon and evening when I fished the Battenkill. One of the big browns briefly followed my streamer, but its efforts seemed half-hearted. And after about thirty more casts, I knew that fish and any others in the area were not interested. The big browns were unlikely to show themselves—let alone take a fly in midafternoon—with so much traffic on the river. So no matter how hard I fished, the odds were not in my favor. Even though a healthy yellow sally stone fly hatch was erupting on the banks around me, I had no takers. While I respect the lore of the Battenkill, I could never feel deeply connected to a river that is as crowded as this one. I was eager to get out of the crowded valley and get up into the mountains—to fewer people, more space, and more naive native trout.

As in all the New England states, brook trout numbers are healthy in Vermont. The Green Mountain National Forest, the backbone of the state, holds hundreds of small brookie streams. But, as in other states, brook trout are often isolated in higher-elevation headwater streams. That doesn't mean you won't catch brookies in some of Vermont's larger rivers, but they certainly aren't dominant.

As I saw in New York, brown trout in particular have become the trophy fish in Vermont's blue-ribbon streams, including the Battenkill. You can still catch brookies in the Battenkill, but most folks travel there for the big wild browns.

Today's healthy brookie populations are a recent development in Vermont. As in other eastern states, the green forested landscape so familiar today did not exist in nineteenth-century Vermont. Sheep grazed on logged-over and often eroded hills and mountains. A century ago, Vermont looked more like a treeless Iceland than the forested environment of today. Logging led to this observation in the *Biennial Report of the Fish Commissioners of the State of Vermont* (1892): "While the forests are being converted to lumber the streams are filled with sawdust and refuse destructive to fish life." Historical photos showing scenes of barren hills where lush green forests stand today seem surreal. When in Vermont, I often wonder, *How could this be the same place?* But Vermont's landscape was indeed heavily used and often abused.

By the middle of the twentieth century, with many pastures abandoned and logging no longer profitable, the forests regenerated. Watersheds again held and released moisture in a way that minimized extremes, creating conditions that encouraged healthy stream ecosystems and brook trout. So, from their refuges in the highest reaches or most remote areas of the Green Mountains, brook trout recolonized much of their former habitat. Lakes and ponds are exceptions because of competition from such warm-water species as introduced bass and perch, but brookies overall are in good shape in Vermont. I have often passed a stream, quickly pulled over, lobbed a fly over a bridge, and caught a brookie. The experience speaks volumes about the resilience and recovery of Vermont's environment.

I headed north and east from Manchester to fish Michigan Brook, the first brook trout stream I fished many years ago. It is a typical smallish tributary of the White River in central Vermont. I am drawn to this stream not only because of my history with brook trout but because of the history of the area. In its lower reaches, where I caught wild rainbows, browns, and brookies, a small settlement existed a century ago. Today markers identify the location of a country store, several houses, and a blacksmith shop that once hummed with life. Stone walls, testaments to people who carved a home from the wilderness, snake through the mature forest today. As I did when spotting old orchards in national forests, I imagined the architects of those walls never envisioned they would be abandoned to revitalized wilderness. Surely they thought their clearing of forested land and collection of stones would result in some sort of permanent presence. How strange to see those walls among mature oaks, white pine, and hemlock.

At first sight, these two worlds—human made and natural—do not make sense or appear to belong together. Some may even lament the presence of the walls because they represent previous human settlement and the destruction of the area's original forests. But I feel they add mystery to the landscape and offer a window into its past. How different this area must have looked when the sun reached the ground without being blocked by the forest canopy.

Although no strict social order exists when these trout species share Michigan Brook's lower reaches—I have caught brookies next to browns—as I moved to higher ground, I found more and more brookies until they became the only trout species present in the frigid high-elevation section. The brookies had the high spruce-dominated, acidy, icy waters all to themselves.

I have had fifty-fish days on Michigan Brook during my numerous visits, but I failed miserably on that trip. So much so that I became impatient. I found myself grumbling, "I know this stream better than any other," and "I have traveled a long way to get here. Take my damn fly!" As silly as it sounds, I guess I believe that, because I know a stream and work hard to visit it, it owes me fish. In the distance at that moment, I heard my old friend the hermit thrush calling from the understory. Its warbling reminded me why I hike up those streams, battling bugs and bruised shins to catch six-inch fish. I do it because of the beauty of the stream, the surrounding forest, the wild fish, and the challenge of placing that fly in that tiny pool in exactly spot where a fish hovers in the current. I needed to remind myself that to catch a fish—and to see it rise from the black pool and take my fly—is just a bonus. The desire to simply catch fish is the source of my suffering when I become impatient. But the Second Noble Truth of Buddhism—the origin of suffering is desire or attachment—is difficult to remember when hours pass without seeing a fish.

Beyond angling, Michigan Brook exemplifies the recreational value of national forests. I usually park just inside the boundary and hike and fish all day without reaching the source of the stream. Thus, I associate mystery, or a sense of endless wilderness (no matter how naive that sounds), with that small stream. As I fish upslope, I never come across NO TRESPASSING signs, I rarely find garbage, and I hardly ever encounter another person. I do share the stream with countless birds, salamanders, and the occasional porcupine that gnaws tree bark at ground level. Best of all, the Green and White Mountain National Forests have hundreds, if not thousands, of similar streams. The large number is the only reason I share the name of this specific stream. I am not really worried about crowds of anglers rushing out to drain its pools of small trout—it really isn't that spectacular in terms of the size of fish. Michigan Brook is just one of many streams that provide the opportunity to get out in nature, but it has come to encapsulate what I love about brook trout fishing.

While I associate Vermont with progressive environmental regulations, its policies for trout management and trout activism are fairly tepid compared with the efforts in the other states I have visited. It is not that anglers or state and federal biologists do not care about brookie resources here, but the state has no shortage of brook trout streams, so perhaps officials have little or no motivation to set strict regulations or ask the legislature for no-take laws. Also, Vermont has many fewer commonly known ponds and rivers with trophy-sized brook trout than Maine has, for example. Most Vermont streams containing brook trout are small freestone mountain streams that usually are home to only small fish. Vermont has undertaken no stream-restoration efforts that remove nonnative trout, and the bag limit on brookies in most streams is an overly generous twelve fish, with no restrictions on size. So Vermont is not perfect, a statement difficult to write because of my deep affection for the state.

I saw an easy streamside parking area too good to pass up, but discarded appliances sat in the middle of the channel of the otherwise picturesque stream. The array was odd for a trout stream. But the rusting stove and washing machine provided in-stream cover for brook trout. Instead of boulders, I fished around old appliances. I thought about moving on, but I was already in my waders, and I was curious to see if trout actually inhabited that stream. I was initially outraged—how could anyone disrupt my vision of what Vermont should be? But even in that stream, I caught brookies. The experience called into question many of my assumptions about brook trout and about Vermont as a perfect and progressive environmental landscape. I wondered how the brook trout, the icon of pristine waters, could live under a rusting stove. I laughed at the situation and myself for my preconceived notions. Perhaps brookies don't care about aesthetics as much as I do. The dumping is more exception than the rule in Vermont, a place I still hold up as unique, even ideal. But that perception is now tempered by the memory of appliance-loving brookies.

It remains difficult for me to believe Vermont has made no plans to more intensively manage a few brookie-only streams with strict tackle and bag limits. During one visit, I asked Dan McKinley, a fisheries biologist with the Green Mountain National Forest, what he thought about a brook trout–only stream. He told me: "Yes, it would be really interesting to restore and protect a stream, to measure how brook trout might respond. But it has never been seriously proposed, because we have plenty of brookie streams, although most are mixed with the other trout species." When I asked him about restricting take on certain streams to protect bigger fish, McKinley laughed and told me, "That won't happen, because the worm crowd would be up in arms about it. Sure, this is Vermont, but there are a lot of people out there that fish to eat."

While the state has not strictly protected any brook trout streams, the U.S. Forest Service has implemented a chop-and-drop program, similar to the State of South Carolina's, to increase the amount of large woody debris in streams. Because of the intense logging of the past, it could take decades or even centuries for forests to become complex structurally and old enough to deliver consistently large timber to streams. McKinley showed me a few streams where he has dropped mature streamside trees into and across channels. The results were impressive, with deep pools and dark water, the weir clogged with rotting, insect-loving vegetation. Midges and small white mayflies swarmed above a couple of pools on the humid day I visited. The fish and the food chain on which they depend had plenty of cover.

In addition to improving in-stream habitat, Vermont is expanding stream connectivity by creating more fish-friendly culverts in streams. Impassible culverts, identical to what I saw under the rail line in West Virginia, often were suspended above the actual stream channel, which makes them impassable for both fish and any other aquatic organisms that move through a watershed. And, given the extensive road networks that exist in national forests, connectivity has been interrupted in literally thousands of locations.

Creating a fish-friendly culvert is not complicated; it is often a matter of size and placement (and money for reconstruction). Corrugated pipes often are not large enough to ensure appropriate water flow for fish, or they are placed too high above the channel, preventing fish from moving upstream to spawn. Traditional passages also have little cover on their floors, making them a sort of a barren "no fish land." Placing the culvert deep enough to capture part of the channel flow and putting stones of appropriate size on its floor mean fish and other creatures are more likely to actually use it. Of course, this more natural passageway allows nonnative fish to move as well, allowing rainbows and browns to expand their range.

Redesigning and replacing large numbers of culverts is expensive and labor intensive. Building them often means closing roads for the duration of the project, inconveniencing tourists and locals. But in the long run, all will benefit. While these redesigned culverts have not been designed exclusively for brookies, they will be a primary beneficiary within both the Green and White Mountain National Forests.

In neighboring New Hampshire, nearly identical restoration programs exist in the White Mountain National Forest, which makes perfect sense because of the proximity and environmental similarities of the two states. Under the guidance of Mark Prout, a fisheries biologist, the U.S. Forest Service in New Hampshire has initiated both chop-and-drop and culvert replacement programs. The White Mountains share cultural and land-use histories with the Green Mountains, and brook trout populations face similar challenges regarding connectivity and quality of habitat. In my daylong tour of the forest with Prout one July, he was especially enthused about adding more woody debris to streams. "Wood has multiple roles in these mountain streams," he told me, "storing sediment, creating pools, providing cover in pools, roughening up the stream and floodplain to dissipate flood energy and store more water longer in the watershed." The storage aspect is important, because the more water a watershed stores, including in stream pools, the more water there is for brook trout and other creatures. This is especially critical as summer arrives, when temperatures increase and rainfall decreases. Like the Green Mountains, the White Mountains have no shortage of brookie streams, but these projects vastly improve their long-term viability.

The streams in the White Mountains of New Hampshire, especially their eastern slopes along the Maine border, feel like home water to me. And within this region, the Wild River remains my favorite. I have spent a great deal of time fishing in both its lower section, just before it reaches the Androscoggin River at Gilead, Maine, and its headwaters in the heart of the White Mountain National Forest. The Wild offers a little bit of everything: wide open river and deep holes in the lower section with big trout, including brookies, and classic small-stream fishing at higher elevations. I usually stop first in the lower section to catch a few nice brook and rainbow trout by dragging streamers through deep holes, then work my way up as far as I can before I run out of daylight. The lower section, with its huge boulders and deep pools, reminds me of classic trout water you might see out West around Yellowstone, where you can throw flies as far as skill and wind allow.

I love the Wild because, even though its lower section is under a lot of pressure from the national forest campgrounds and roads, I consistently catch fish. Higher up, the pressure from people subsides, and I rarely see anyone fishing the small streams. The upper elevation streams are full of small brook trout, with twenty-to-forty-fish days fairly common. I sometimes pack a lunch and hike up the Wild's tributaries all day without ever seeing another person. I fish pool to pool, pulling a brookie or two out of every one on good days. They rarely exceed six inches, but even a six-inch fish that crushes a dry fly is a thrilling moment for a brookie angler. Based on these experiences, I want to reiterate my support for the strategy followed by Prout and the national forest service managers: using downed timber to create deep pools, a characteristic of the most productive headwater streams. When the pools disappear on certain stretches, so do the fish.

Back in the Green Mountains of Vermont, I found myself again at the Robert Frost Interpretive Trail near Ripton in central Vermont. The brookie stream was and probably remains fairly marginal for fishing. I was drawn to the location by the place name, the area's association with Frost, and knowing I was close to, or even on, the same ground he once walked.

Frost's poems so vividly and powerfully capture the raw spirit of rural New England that I linger on his words. They are personal, as was his own relationship with the land and people of New England. When I stay near Royalton (a good central location for angling in the Green Mountains), the Frost Trail area is about as far as I travel in a day, and then I work my way back, one stream after another. That stream always surprises me with its lack of trout, even in fishy-looking spots. But I fish nevertheless, thinking of Frost's poems, especially "Birches"(1916), "Stopping by Woods on a Snowy Evening" (1923), and, of course, "The Mountain" (1914), in which he wrote:

> Of easy wind and downy flake.
> The woods are lovely, dark and deep.

I can see Frost in this landscape, stopped with his horse and carriage while heavy snow makes a slight soothing sound as it coats the trees and fields. And the stone walls, many of which are now crumbling, are perhaps the same ones that inspired "Mending Wall" (1914), a poem in which he analyzes the annual ritual of reconstructing walls on his land after a hard winter. I feel as though I know more about this landscape, including what it was like in Frost's time.

After a morning of dreaming about the poetics of place, it was time to move on, to move down into the White River Valley and find more brookies. But I found myself spending more time looking at the landscape than fishing. I had already caught a few brookies, so I did not feel the need to stop at every stream. Instead, I stopped at a weathered farmhouse to buy some maple syrup. Signs advertising maple syrup are ubiquitous in Vermont, as are the thin black tubes that in early spring transport sap to collecting buckets. They run like dark spider webs throughout the forests, and one must practice a sort of limbo move while hiking in some areas. The couple on this farm belonged to a new generation of farmers trying to recapture a simpler time. They told me they also raised goats for meat and cheese: "We are cobbling an income together." They loved Vermont and wanted to raise children there. I was envious. They would stay and would see the fall colors, the snow, and the gray muddy thaw of spring. I moved on, imagining what this valley looks like at peak leaf season. I bought some syrup and goat cheese, my lunch that day, and wished them well, hopeful that, on my next visit, they would still be there.

My final destination was high in the Green Mountains, where I fished until dark and then camped. A light hammock tent, one literally hung between two trees, a single Long Trail IPA, some water, and a few snacks were all I carried. I slept in the forest and listened to the night sounds above the din of the stream. I chose a stream in the central Green Mountains, a stream dominated by brookies. I am amazed how unpredictable nature is. Why would one stream be populated only by brook trout, while others, nearby and in the same watershed, be full of rainbows? There is really no obvious answer, for there are no natural barriers and no efforts to remove nonnatives. This was a small stream from start to finish, but I have certainly caught rainbows in smaller trickles.

It was a perfect afternoon—sunny and clear—and the water was high and slightly off-color, a result of recent summer storms, but not so mocha that a dry fly would go unnoticed. The water flowed over and around boulders and downed trees and under spruce snags. The sunlight twinkled through the dense canopy. If I were claustrophobic, it would have been easy to feel hemmed in. As I climbed, I glanced into the forest canopy and noticed a silent barred owl watching me from a broken white pine. I have no idea why I looked in that direction—my focus is usually squarely on the stream. I wondered again what else had I missed because I was too zeroed in on the end of my line. I moved on and out of sight of the bark-colored bird. I fished fast, because I wanted to get high up in the headwaters. I threw a fly only to the best-looking pools and even then made only a few casts. I caught brookies here and there, usually below undercut banks or in the dark slack water behind downed trees.

Even by fishing fast, I did not reach the source of that stream. *Another mystery goes unsolved*, I thought. I am not sure I will ever fish fast enough to find a stream's source. It was a good effort, though. And maybe I did not really want to find the true source of any stream, to reveal the mystery about that water. Thinking again of Peter Matthiessen's *The Snow Leopard* (1978), I asked myself, "Have you seen the snow leopard? And the answer is no! Isn't that wonderful?" Mystery and not finding everything, in this case the beginning of the stream, can be a good thing.

I have always felt that is it good to walk away from a stream with some ambiguity, that is, without catching the most fish or biggest fish of my life or finding the stream's source. If I truly knew that the last brook trout I caught was the biggest, the most beautiful, the hardest fighting— or that I would never catch more fish than I had on a particular day—would I really ever so enjoy angling again? I doubt it, because a large part of the fun in fishing is the surprise. The mystery of what lies in the next pool or around the next bend would be gone, and the quest would be over. I doubt that I am alone in this belief. That is why the fish that got away was always the biggest and most memorable. I find I do not retell the story of the pretty eight-inch chunky trout I caught but instead recall the fifteen-inch mystery fish that broke off under a log. I have heard similar sentiments from many anglers. We all want something to reminisce about and to look forward to the next time we fish, our own Moby Dick.

A hermit thrush called in the distance as the shadows stretched to their farthest point. As the forest grew dark under the canopy, I called it a day on a small ridge above the stream. I secured my beer to let it cool among the rocks in the stream while I hung my hammock. I looked down toward the stream and saw a huge stone fly flittering with the current. As dusk turned to dark and before my eyes adjusted, I was amazed at how pure black the forest appeared. There were no stars, just a shadowy moon and wind whistling through the hemlock and spruce. Thank God for the wind, which kept the bugs few and far between. An owl—maybe the one I saw earlier—screamed in the distance. After enjoying my precious cold beer, I flung myself into my hammock and was lulled to sleep by the muffled rumble of the stream. I slept well, dreaming of big brookies and the deep dark woods that sheltered me.

CHAPTER 14

Maine

AN EIGHTEEN-INCH brook trout slid out of my hands and into the deep black pond in northern Maine. Its cold belly sagged around my fingers just before I let it go. Its weight carried it down like a copper torpedo. The fish remained still for just a moment, then, after a quick flick of its tail, disappeared into the depths. Small bubbles were all that remained of my trophy. I looked around, breathing in the sweet smell of balsam, the scent of Christmas, and felt quite happy with myself after losing a few other big fish earlier in the day. With my next cast, my thoughts turned greedy—I was hoping an even bigger trout might be down there.

Within fly-fishing circles, Maine has the reputation of being the center of brook trout angling in the eastern United States. There, brookies are not limited to tiny headwater streams but are found in all waterways, from the smallest trickle to the largest rivers, as well as many lakes and ponds. The fish are a major draw. Few people travel to Georgia or even Pennsylvania exclusively to fish for brook trout, but fly anglers from all over the world travel to Maine for its brookies, the fish on which the

storied sporting-camp traditions of the North Woods and Rangeley Lakes were based.

Maine has more than a thousand glacial lakes and ponds that contain brook trout. Of these, 645 are self-sustaining wild populations, and 305 have never been stocked. This last statistic is important, because it means many of these lakes and ponds have been home to brookies since the end of the last ice age, and they contain genetic strains found nowhere else on the planet. Brook trout also live in more than twenty-two thousand miles of Maine's streams and rivers, and the vast majority are wild fish. This estimate is on the low end, because many smaller streams in Maine have not been surveyed at all or only anecdotally. As in other states, brookie numbers have declined since the 1800s, but they have been wiped out in fewer than one percent of Maine's river and stream watersheds, which remain in far better shape than those in any other eastern state. In addition to serving as home to my family, Maine was the brook trout Shangri-La on my trout checklist.

Within Maine, the center of brook trout angling is the North Woods. Located in the far northwestern section of the state, the North Woods contains more than 3.5 million acres of commercial forestland, most of it privately owned. The landscape is part of the New England–Acadian forest ecoregion of mixed northern hardwoods and conifers. Common tree species include American beech, balsam fir, black spruce, eastern white pine, northern white cedar, paper birch, quaking aspen, red spruce, white spruce, and yellow birch. Biogeographically, the North Woods serves as a gateway to the great taiga forests found in Canada. The area is home not only to boreal tree species but also northern animals such as crossbills, spruce grouse, moose, and, until the early twentieth century, woodland caribou and gray wolves.

I spent my first time in the North Woods at Libby Camps, a sporting camp that has been in the Libby family since the 1890s. Although I have lived off and on in Maine since 1999, this was my first true journey into the heart of the North Woods. I had dabbled around the edges but had never stayed in the heart of it. The North Woods is largely depopulated, crisscrossed by rough logging roads, so it is not the kind of place you casually visit, especially with small children.

The North Woods is almost mythical. On a map, it is terra incognita, and the only discernable human imprint is long straight logging roads, most labeled as private or with numbers whose meaning is unintelligible to the outsider. As a geographer, I have always been fascinated with blank spots on maps. When I peer down on maps showing the almost-empty Mosquito Coast or Canadian taiga, the lack of place names makes the areas even more appealing. And northern Maine is similarly intriguing to me, albeit on a smaller scale. I told myself that surely a blank spot

means big trout. Most world religions have important destinations for pilgrims; Islam has Mecca, Catholicism has Rome. Brook trout pilgrims have Maine's North Woods. Catching that eighteen-inch trout truly felt like the end of my quest. I felt at home on many levels.

The North Woods is the last and biggest wilderness area in New England, rivaled in the Northeast only by Adirondack Park in New York. And though the Adirondack region is impressive, Maine's North Woods has far fewer people and towns and certainly fewer paved roads. It is a working forest, mostly an unprotected area owned largely by timber and paper companies. Yet the area maintains healthy populations of game and nongame species and is visited by thousands of sportsmen and -women every year to hunt for bear and partridge, fish for brookies and landlocked salmon, and paddle secluded waterways such as the Allagash River.

The rough roads that snake through the North Woods have a reputation as unwelcoming because logging trucks literally and figuratively own them. Speeding trucks bearing freshly cut logs have been known to run cars off the road. I wondered if my car ever would be the same after scraping the oil pan on one exposed rock after another as I navigated the rutted roads, but my drive was uneventful. It was Sunday, and I saw no other vehicles, let alone speeding loggers, during my hour-plus drive. I even stopped at a few small bridges to check out the streams. Roadside vines and shrubs were coated with a thick layer of chalky dust, and no trout were visible from the road. Few places in the eastern United States are as quiet and bereft of human noise as these woods. In the heat of that sunny July afternoon, the area was almost silent—no trucks, no cars, no planes, just roadside weeds and, beyond that, wet spruce forests. And from that wet spruce I heard the faint yet familiar call of the hermit thrush, a comforting sound because I associate it with a brook trout landscape.

As I pulled into camp, I found it was everything I had imagined: log cabins surrounding a large dining hall, all overlooking huge Millinocket Lake. The floatplane was waiting at the dock to take me to my brookie nirvana. It looked and felt like a tiny isolated settlement from another era, which in many ways it is, given that it is roughly thirty miles as the crow flies from Interstate 95. Make no mistake about it: Libby Camps is a lakeside fishing and hunting camp near Ashland; it is not a resort. As I got out of my car, two of Libby's Brittany spaniels barked with suspicion. Their presence reminded me that, in a few months, partridge, not brook trout, would be on the minds of guests.

Camps in the North Woods are unique because people hunt and fish on largely private lands owned by timber and paper companies. This requires a curious symbiotic relationship between those who need an intact environment and those who sometimes completely alter it. Private individuals and corporations own more than 90 percent of Maine's northern forests, which means the land isn't protected in a traditional sense. As Libby Camps has been in business for more than a century, this arrangement has worked for its owners, but the timber companies and their practices are not uncontroversial. One Maine group, Restore: The North Woods, sought to create a national park on timberland that surrounds the 200,000-acre Baxter State Park, located to the south of Libby Camps. The original champion of this initiative was Roxanne Quimby, the cofounder of the Burt's Bees personal care

product empire who has purchased tens of thousands of acres in the region. The idea has not gained popularity in many northern Maine communities, with the opposition aptly summed up in a 2001 letter to the editor of the *Bangor Daily News* from Eugene J. Conlogue, the town manager for Millinocket: "Message from northern Maine to Quimby: Leave us and our way of life alone. We don't want your park or your unfortunate, inaccurate and unfair depictions of the hard-working people of our part of the state." Ten years later, the Maine State Legislature overwhelmingly approved a resolution expressing opposition to a national park. Then-president Barack Obama nonetheless designated the land donated by Quimby as a new national monument in August 2016. The Katahdin Woods and Waters National Monument contains more than eighty-seven thousand acres and initially was supported by a multimillion-dollar endowment, also donated by Quimby.

Rural Maine, especially the northern counties, is a region with historic ties to the timber industry, big-game hunting, and potato production. Any proposal that seeks to curb these pursuits—especially timber—will meet great resistance. Local residents have often resisted creation of parks in the United States. For example, local residents and politicians opposed the creation of Great Smoky Mountains and Shenandoah National Parks and the complete removal of local residents by way of eminent domain. But I thank God that those parks were created, even in the face of some local opposition.

Although I am an ardent conservationist, I recognize that loggers can use methods that do not destroy the environment. I also recognize that disturbance, even large-scale disturbance, can benefit species that depend on early (that is, young) successional or disturbed environments (a successional environment is one whose mix of species and habitat changes, for example, are a result of fire). The Canadian lynx, endangered in Maine, is an example of a species that depends on early successional dense spruce forests, because these environments hold its favorite meal, the snowshoe hare. Of course, many other species, such as the American martin, do not benefit from the logging of mature forests, so it is a two-way street. When I flew over the North Woods in the camp's pontoon plane, owned by the camp, where new and old forests met was obvious. Mature dark-green spruce trees abutted the regenerating forest, which was almost lime green. The edges no doubt provided habitat for a variety of birds and mammals. So, again, disturbance is easy to categorize as good or bad, but which it is it depends on the needs of individual species. Nature, even when protected in a park or reserve, experiences disturbance, sometimes on a large scale.

Even recognizing the role of disturbance in maintaining biological diversity, I truly believe the new national monument represents a great opportunity for a part of Maine whose economy and population have been left behind for decades. Middle-class mill jobs and even timbering jobs (reduced by automation) are not ever coming back to northern Maine. The new protected area, also championed by Quimby's son, Lucas St. Clair, represents an attraction around which other sustainable outdoor industries can develop. Given the shuttered storefronts and high unemployment in the region, opposition to the gift always seemed to me to be incredibly shortsighted, if not economically masochistic.

Automation is not the only dramatic change the timber industry in northern Maine has undergone in recent decades. The industry has moved from one dominated by family-owned timber operations managed for the long haul to industrial forestry corporations that have no personal connection to or history on the land. Since the 1990s, millions of acres of the North Woods have changed hands; investment companies bought a large percentage of that property. Powerful corporate logging companies are able to violate the law time and time again, often by cutting too close to pond buffers, paying a relatively small fine, and proceeding with the cutting. They are not beholden to their neighbors and local community. This seems to be the trend in northern Maine. According to the Appalachian Mountain Club, since 2000 clear-cuts in Maine have become larger and rotations between harvests less frequent. This trend will not sustain healthy populations of brook trout, which depend on intact riparian forests and sheltered feeder streams.

Economies of scale have caught up with many family-owned timber interests, which now see rapidly declining profits in a global economy. Large tracts have become concentrated in the hands of increasingly fewer industrial logging interests. Not all logging is incompatible with sustainable management of brookies, or any other species for that matter, but the increasing ubiquity and power of large-scale industrial logging are alarming.

To further complicate matters in the North Woods, timber companies are developing and have initiated plans to turn large tracts into expansive and expensive residential housing. The Plum Creek development in the Moosehead Lake region is the most prominent example. Plum Creek, a

Seattle-based developer purchased by Weyerhauser in 2016 (making Weyerhauser the largest private landowner in the country), owned about 900,000 acres in Maine, including land that contains hundreds of wild brook trout ponds. The initial plan for the acreage called for more than nine hundred building sites, two large resorts, and an undetermined number of related businesses. The sheer number of roads alone required to construct and support this development would have unleashed unprecedented environmental change. As early as 2007, the Maine Department of Inland Fisheries and Wildlife stated: "There will be a decline in the quality of the fishing experience for anglers in the Moosehead Lake Region." And state biologists predicted a fourfold increase in fishing pressure. Fortunately, the original plan has changed considerably, scaling back much of the proposed construction after review and critiques by Maine's Land Use Regulation Commission (and widespread public outcry at the original scale of the proposal). For example, some of lodges and other infrastructure will now be located closer to existing towns to minimize sprawl and better connect local communities with employment opportunities. Weyerhauser has placed thousands of acres in conservation easements in return for the approval of Maine's Land Use Regulation Commission. So while the plan won't be without environmental impact—no large-scale development project ever is—hundreds of thousands of acres of forestland (and trout ponds) now are permanently protected by conservation easements as the Maine commission stipulated. Although the conservation easements are in place, no construction had started as of mid-2022.

I arrived at Libby Camps in July, so I missed many of the insect hatches and dry-fly fishing. But while the purist in me loves nothing more than watching a trout rise to a fly, I also wanted to catch fish. And big fish are more often caught below the surface, where they are not exposed to predators and where water temperatures are more constant. During my brook trout pilgrimage, I learned that anglers cannot impose their will on nature. No matter how much we might like fishing a specific dry fly, we have to deal with the opportunities nature provides. As the ponds heat up, fish seek deep cool water or the thermal refuges springs create.

Rick Young, the guide I had booked through Libby Camps, sensed my disappointment with sinking lines and firmly told me, "I get a lot of sports who initially insist on using dry flies. But I have to ask them: Do you actually want to catch fish?" When a guide presents the situation in such direct terms, you do what the guide says. Young noted that he has a few clients who insist on fishing their own way. "Oh, yeah, some ignore my advice," he says. "That's up to them, but it can get pretty boring after a couple of fishless days." During a long first day of seven, I caught more than twenty brookies. A few green drake mayflies made an appearance late in the afternoon, but given their numbers, they were just teases that failed to draw many fish to the surface.

Dinner at Libby Camps is family style, with heaping servings of meatloaf, rolls, green beans, salad, and potatoes. Because I travel alone, it's good to compare notes with other anglers in the evening. We have to talk about and compare our fish in order to validate our success. One young angler caught a huge brookie, twenty inches or so, in a nearby pond that sits right next to the main road. Not the sort of place you would expect an old fish, given the easy

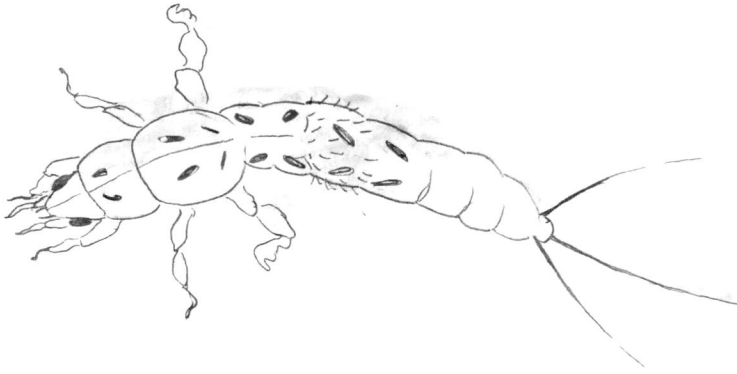

Eastern green drake mayfly nymph (*Ephemera guttulata*).
Drawing by Frederick Steinberg.

access to the pond, but maybe a lot of people overlooked the pond because everyone traveling past expects it has been fished hard. I sought out the most remote locations, either flying in the camps' plane or driving in Young's truck for what seemed like hours on rough roads while that teenager walked a few hundred yards and scored the fish of a lifetime. There is a lesson here. But in the young angler's desire to get one more photograph, the fish died in the heat. (Memories are often not enough; photos somehow make the catch more real.) Brookies, so attuned to specific temperatures and oxygen levels, are quite fragile during the hottest parts of the summer. Ellen Libby cooked the fish that evening, so at least it wasn't wasted.

In addition to eating, drinking, and bragging, anglers spend their evenings outlining their fishing plans for the next day. More experienced fishers subtly jockey for the best ponds, or at least those that recently fished the best

or produced memorable fish. Rumors fly around the lodge about such-and-such pond, and everyone pesters their guide about getting there: "I heard someone caught a six-pound fish last week on Spruce Lake, so can we go there tomorrow?" Young laughed at this, saying, "There are so many ponds that receive minimal pressure that any of them can and do produce big brookies." But we all want the lifetime fish—that five-pound-plus brookie found nowhere else in the United States. We become infected with big brookie fever. All of us also are there for just a short visit, so everyone feels the same pressure, the sense that the fishing clock is ticking away.

This fever causes anglers to go through an amazing metamorphosis. At first, we deem the ten- to twelve-inch brookies great, bigger than any brookie we have ever caught. Next, we declare that the fifteen-inch fish is amazing and claim, "I never expected to catch such big and beautiful fish." Then, if we are lucky, we get to commune with the eighteen- to twenty-inch fish, the real lifetime brookies. After catching, hooking, or at least hearing about these trophies, we can't release twelve-inch fish fast enough. Now we toss back without a glance the same fish that we had once insisted on photographing. It is a greedy addictive cycle.

I try, unsuccessfully, to avoid this trap of wanting and possessing the fish of a lifetime and then being disappointed when I don't catch such a fish. I never thought I would be disappointed with, or want a brookie that is longer than, fifteen inches, but that is what happens when my expectations run amok. Talk about not living in the moment. The fish of a lifetime happens once, and we do not realize it when it does happen. When I really think about it, will I be able to identify my fish of a lifetime only on my deathbed? So how or why do we desire such a thing? As we refine our skills and catch bigger fish, suppressing our expectations becomes increasingly more difficult. This cycle of desire is also difficult to avoid because we can't know when or if we will return and be in a position to catch a trophy fish. Everyone at the dinner table claimed they would be back the next summer, but no one could be sure of it. Next year, a new glossy brochure for another location will catch our eye. I decided then and there that one day one or both of my sons will accompany me to a fishing-only lodge.

The next morning, we anchored the canoe as the sun crept up over the trees, and Young warned me that the lake we were fishing was home to a trout-stealing loon. I found this hard to believe, because loons always seem so shy and reclusive, more frequently heard than seen and then usually at dusk. My reaction was excitement: "I'd love to see a loon up close." My next thought was, *What sort of gullible fool does he take me for?* I soon I realized it was not a story he had made up for tourists. After catching and releasing a few nice bronze brookies in the twelve- to fourteen-inch range, a loon appeared and slowly circled the canoe. My guide said, "I told you so."

I still had my doubts that it would approach the canoe at close range. But as I fought another nice brookie, a black prehistoric-looking shape with a harpoon-like bill shot through the water just below the canoe. It was huge and frankly a little frightening. I never expected this solitary bird, whose call is synonymous with Maine and the North Woods, to be so eerie looking under the water. It was a ghostly, unexpected shape, more like a plesiosaurus

than a bird. After a few more sorties by the loon under and around the canoe, the brookies scattered like buckshot. We moved. The loon made a few more passes later that morning but finally disappeared as the fishing slowed. I will never hear that distinctive call of the loon the same way again.

Thieving loons aside, the North Woods is a remarkable place. Only a half-day's drive north from Portland, it might as well be another country. In my travels along the east coast and in the Appalachian Mountains, I visited few places where I wasn't surrounded by people and cities. With few exceptions, brookie fishing in the East is the only type of fly-fishing that doesn't mean stepping on other people or picking up their beer cans. Driving north on I-95 past Bangor, the landscape gets quieter, with fewer towns, people, and cars. The farther north one travels, the quieter it gets. Mount Katahdin (5,269 feet), the northern terminus of the Appalachian Trail, glows in the distance: green below, with bald gray rock up high. The North Woods is different: not pristine, by any stretch of the imagination but definitely wild. During my stay at Libby Camps, I saw plenty of moose, bear tracks, eagles, spruce grouse, one too many loons, and dozens of big brookies. To borrow a phrase from Maine's Office of Tourism, that is "the way life should be."

In many respects, the threats to Maine's brookie populations are the same as those in many states. Rainbows and browns and nonnative, warm-water fish species such as bass pose serious threats in many waterways. The threat posed by bass is ironic, given Maine's northern location. Large- and smallmouth bass, pike, and muskie, among many others, have had devastating impacts on native fish, especially brookies. Remote ponds in which brookies have evolved without predators are especially vulnerable. Sportsmen (in name only) have introduced many of these fish illegally because they want to catch bass instead of trout. Merry Gallagher, a Maine state biologist, told me about an individual who was caught with a bucket full of live crappie that he intended to release in a brookie pond. If those fish survived, they could have devastated the native fish population through predation.

Examples abound of illegal introductions of fish, including largemouth and smallmouth bass, minnows, and sunfish, that weren't caught in the act. While transporting and dumping live fish is illegal, only a handful of wardens safeguard Maine's thousands of lakes, ponds, rivers, and streams, so skirting the law is not difficult. Perhaps the most tragic and still unfolding example involving brook trout and introduced warm-water species occurred in the famed Rapid River in western Maine. At only a little more than three miles long, the Rapid is a short but rough and beautiful river that historically supported big brookies. Until the latter half of the twentieth century, anglers could often pull five-pound fish measuring more than eighteen inches from its deep pools and torrents. Why a roughly three-mile river would grow such big fish has never been clear, but some speculate that the abundant baitfish—in this case, smelt—provide the ecological foundation for big fish. The Rapid still has big brookies, but their numbers are but a fraction of what they were a few decades ago.

A major reason for the decline of brookies in the Rapid was the illegal introduction of smallmouth bass at adjoining Umbagog Lake on the New Hampshire border during the mid-1980s. In less than a decade, smallmouth bass moved from the lake into the Rapid, where they began feeding on brook trout fry, displacing brookies from prime holding areas within the river. Smallmouth bass are aggressive predators, so their appetites seriously threaten the ecological berth of brookies in the Rapid. As brookie numbers have declined, landlocked salmon have moved into this void and further marginalized the Rapid's native brookies. This niche replacement is probably just one of the more visible examples of a variety of ecological problems that the brookie and other native organisms face in this aquatic ecosystem.

Based on its historical reputation, the Rapid, like the Battenkill and Beaverkill, is a "must fish" river. I met the Maine state biologists Dave Boucher and Forrest Bonney at a cabin they used during stream-survey work near the riverbank. (Boucher died unexpectedly in 2016, while Bonney has since happily retired and started a second career making furniture.) Over beer and grilled hamburgers, we spent the evening talking about brookies. Bonney had long been an advocate and recognized leader in conserving Maine's brook trout populations. His four decades of research for the state helped create more stringent take and tackle regulations for Maine's fabled brookie ponds and culminated in a book, *Squaretails: Biology and Management of Maine's Brook Trout* (2007). Boucher also had impressive credentials. He worked on cold-water fisheries issues for many years and wrote several reports on the Rapid. Both lamented the decline of the river, but they also assured me that brookies were not going to disappear entirely. "We can at least knock back the bass numbers with altering the water releases from the dam," Boucher said. "Certainly, the smallmouths are now a permanent presence in the river, but we can make sure they don't completely displace the brookies."

Bonney agreed with Boucher's assessment. They based their opinions on a 2007 report by the Rapid River Coalition, to which Boucher contributed, that says that increasing flows can reduce overall bass numbers, or at least spawning success. Intensifying the flow from the dam during the bass spawning periods in early summer literally blows out their nests. "Nonnative fish are our biggest problem here in Maine," Boucher added. "We have pretty good laws protecting the resource, but we still have people moving these fish around. It's nearly impossible to prevent if someone really wants to move fish."

According to Bonney, the culprits include baitfish. "It's important that the public understand that shiners and smelt can have just as big of an impact on trout ponds as bass. And, in recognizing that fact, we need to restrict the use of baitfish on ponds. They are just too valuable." Both men also were adamantly against even the idea of stocking hatchery trout in or around the wild brookie ponds, even in private ponds. Many of these ponds have no stocking records at all, and both men felt it was crucial they stay that way. "If there was ever an issue we need to draw a line in the sand about," Bonney said, "it is the stocking issue."

While the manipulation of flows can control the impact of smallmouths, little can be done to eliminate them. The Rapid is not a small confined stream from which fish can be removed by using piscicides or electroshocking. The Rapid is a powerful river connected to a lake that acts as a source for smallmouths. The effects of smallmouth bass were clear the next day, when I fished the Rapid. While my visit was short—one day of fishing—I saw small hatches

March Brown mayfly (*Maccaffertium vicarium*).
Drawing by Frederick Steinberg.

of insects throughout the day, mainly small dark mayflies and tan caddis flies, so the fishing was consistent. And I caught plenty of bass, a couple of salmon (also introduced), and fewer brookies, as I slowly fished up a long stretch of the Rapid.

This, of course, could change, depending on location and fishing conditions, but I pulled in plenty of six- to ten-inch smallies. I killed them all. I did not like doing that, but they have no place in the Rapid. I do not like tossing any fish on the bank. I practice catch-and-release almost exclusively. I do not kill rainbows or browns when I catch them in a brookie stream, as frustrating as that is when I find them in some remote trickle. But the Rapid is different, such an important brookie stream that the State of Maine even posted signs urging anglers to remove all bass. I do not fault the bass for being there. They are doing what any organism is supposed to do: survive and multiply. But something has to give, and unfortunately for the smallies, they are it.

Sadly, Maine has many other examples of destructive invasives, including the Québec government's introduction of muskie into the St. John River in 1970 to create a new sportfishing option. The muskies have been so successful in the main stem of the river that a once-thriving brook trout population is gone forever. The muskie ate all the trout as the trout congregated in large numbers in spring holes during the hot summer months. Muskies today are so common and large, some reaching more than twenty pounds, that the Maine town of Fort Kent, located on the Canadian border, holds an annual International Muskie Fishing Derby. The St. John River was once home to brook trout in the twenty-inch range; it rivaled the Rapid River but no more. And now the muskies threaten to expand their range and decimate the region's remaining brookie strongholds such as the Fish River in northern Maine. Muskies are found literally just below an impassible waterfall near Fort Kent. Eight large lakes, known as the Fish River Chain of Lakes, are located on the Fish River. All these large lakes, as well as many small ponds and interconnecting streams and rivers, contain wild and native brook trout populations and a lot of potential muskie habitat. If muskies become established in the lakes, they eventually will destroy another outstanding brookie fishery.

Many other examples demonstrate the negative impact of invasive species on brook trout and other native species in Maine, ranging in scale from single ponds taken over by voracious smallmouth bass to the almost complete decimation of the Maine blueback trout (*Salvelinus alpinus oquassa*) by introduced rainbow smelt that fed on trout fry. Through the state's intervention, the blueback has come back from the brink of extinction, but its position remains tenuous, because an individual with a bucket of baitfish could always destroy a pond's ecosystem, thereby jeopardizing years of restoration efforts in an instant. Many native brookie populations have evolved in isolation, but once they are eliminated from pond A or watershed B, that is it—thousands of years of natural selection are gone forever. "The threat from nonnatives makes it that much more important to protect and improve brook trout habitat in other areas," Bonney said. "There is always restoration work that needs to get done."

The day I spent on the Rapid left me feeling melancholy. Certainly the scenery was amazing, with white pines and hemlocks converging at the edge of the river with its deep black pools. I saw a bald eagle, its jet-black body and contrasting cloud-white head, float by on rising thermals. And I had the river all to myself, until I reached the large pools just below the Middle Dam, a well-known angling landmark in Maine made famous by the fishing and by the scenery. But strike after strike brought in only a bronzed smallmouth. Had I been fishing farther south, where bass are native, I would have felt satisfaction, but here that catch just did not seem right. Had I fished differently or in other locations, perhaps I would have caught more brookies and not felt like the bass were winning.

As I left the Rapid and drove back to Maine's southern coast, I came away feeling like we had really screwed up one of the best, most productive, native trout rivers on the planet. It was great to be able to say, "I fished the Rapid." But unlike so many other places I fished, I will not dream about my next trip to this beautiful but now highly altered river. I fear the days of five-pound brookies might soon exist only in my photo albums.

For every selfish angler who releases a nonnative and potentially destructive fish into a new environment because the fisher finds the native population unsatisfactory, many more anglers, outdoor enthusiasts, and state officials give time and money and/or dedicate their careers to protect and restore the environment, including brookie streams. Maine has no shortage of such environmental good Samaritans. Sporting and conservation groups—often one and the same—have been instrumental in working to remove dams that block the passageways for migratory fish such as Atlantic salmon and that also isolate brook trout populations. The removal of the Edwards Dam from the Kennebec River in 1999 served as a model for dam removal across the nation, and it really ignited the interest among cold-water conservationists in removing antiquated dams. Today striped bass, alewives, and other native sea-run fish have access to seventeen more miles of river habitat as a result of the dam's removal.

Anglers in Maine have contributed mightily to a variety of low-profile yet critically important activities. These include riparian stream restoration work, fish-friendly culvert construction, educating the public about the importance of cold-water fisheries, holding fly-fishing clinics and camps for kids, raising money for, and making

donations to, cold-water conservation, and lobbying politicians when trout and salmon are at stake. One of the more inspirational activities of Maine trouters is Project Healing Waters, which provides fly-fishing and recreational opportunities for disabled veterans.

A group of volunteers and military veterans happened to be staying at Aldro French's lodge on the Rapid River during my visit. They were just downriver from where I was staying, so I had the opportunity to meet them. These were men and women had paid a great price to protect this nation. Most had suffered severe injuries, yet with the help of French, the lodge owner, and volunteers, most of them from Trout Unlimited, the veterans were able to spend some time in one of the most beautiful places in Maine, learning or relearning to fly-fish. The activity can be so restorative, both physically and mentally, that it makes perfect sense for the veterans. The next day, I took a break from fishing and watched a couple of participants cast their flies below Middle Dam. They were struggling. Several had a hard time delivering their cast more than a few yards. But they kept at it, and volunteers kept encouraging them. After about twenty minutes, one veteran caught a small trout. Shouts went out, and the group shared enthusiastic high fives.

While I have highlighted a few of the significant challenges Maine's brookies face, the state still boasts more brook trout waters than any other eastern state. Certainly, brookies' range has receded during the past few centuries as agriculture, logging, and industrialization transformed the rural Maine landscape. The loss of habitat is especially clear along Maine's southern coast, where most of the state population lives and works. But brookies live on, even in southern Maine, and you can find them if you look closely enough. The angling opportunities also are varied—lakes, coastal rivers, ponds, large rivers, freestone mountain streams, meandering spring-fed streams, and beaver ponds—not just tiny headwater streams as in so many other states.

Even though Maine has many brook trout angling opportunities, I found myself without a true home water in this state. In Vermont, one stream made an indelible impression on me, but in Maine it remains easy to fish for brookies all summer long and not fish the same stream twice. When I spent my summers there, I did have a few favorites, including a spring-fed creek that consistently produces wild twelve-inch brookies. This stream is not far from Portland, but its vegetation-entombed banks are full of ticks and keep the faint of heart away. Or maybe it is another well-known stream in southern Maine in which I consistently caught fourteen-inch browns. Or maybe it was the Wild River, where Frederick caught his first trout on a dry fly. I always imagine the pool around the next bend or the new stream I found in the *Maine Atlas & Gazetteer* will be even better than the previous one. Even with the aid of hindsight I'm not sure which stream was my favorite.

Because of these endless opportunities, I did not panic if I missed a day of fishing when I was in Maine. There was always another day, whereas when I am in Alabama, I have to drive about five hours to reach the closest brookie streams. In the Smoky Mountains, for example, I stay for the weekend and fish during every minute of daylight. I often cast too fast. I often become impatient. I often fish too hard. In the end, I fish sloppily. I know that I am on the clock and need to cover a lot of water before my weekend ends. Not so in Maine. There I got out my state atlas and truly thought about each place I might visit. In Maine, I did not have to fish fast and sloppy, because I could always return another day, or so it seemed at the time. I savored each day instead of looking at my watch. Time really did slow down.

The way life should be. I miss it.

CHAPTER 15

Labrador

OVERCAST PEWTER SKIES greeted me in Wabush, Labrador. It was mid-August, and cool weather comes early at this latitude. There were few hardwoods, so I did not see any colors, but the sky hinted fall. Windswept and gritty, Wabush had the look and feel of an arctic frontier town with rusty pickup trucks, tough-looking men, and quickly constructed corrugated buildings. Wabush is not a resort town. Two large mines on its outskirts were visible on the flight in. The mining operations had left massive gashes in the spruce forest of the Laurentian Plateau. Rust-stained runoff bled from open mining pits into nearby lakes. Iron ore and hematite, not brook trout, were the mainstay of Wabush. Even in remote Labrador, the destruction of humans was clearly visible.

When I started this journey to catch brookies in every state from Georgia to Maine, I assumed I would end my pilgrimage in Maine, because it was the end of my brook trout path every summer when I traveled from Tuscaloosa to Portland during my marriage. But as time passed, it struck me as illogical to stop in Maine at the U.S. border with Canada. Certainly, brook trout recognize no political boundaries. A pilgrimage ends at a significant or even sacred destination, not a line on a map. If I truly wanted to explore native brook trout landscapes and their home waters, I had to follow the fish to the place where they are largest in their range. Nowhere else on Earth do brookies reach the size they do in Labrador.

I was perfectly content to end my journey in Maine, until I saw a YouTube video featuring huge brook trout being hauled out of various Labrador rivers. It was a promotional video for Three Rivers Lodge in northern Labrador. Any promotion features only the biggest and best fish, I knew, but I still was captivated. I tracked down some former guests of the lodge, and they confirmed that I was certain to catch a lifetime brook trout. Next came the conversation with Barb, my wife. She agreed that it made sense to complete the journey in the landscape of giant brookies. "How could you not finish the book there?" she asked.

The next thing I knew, I was booking a flight from Portland to Wabush and paying more money for a fishing trip than I ever could have imagined. In the end, the trip cost more than $5,000, which tested the limits of my obsession. Barb, though supportive, also pointed out that, for the price, I could probably fish just about anywhere in the world. I retorted that few places in the world have ten-pound brook trout. Desire and the lure of the big fish won. Barb was not angry about the cost; she understood my motivation and the overall book project. But she made it clear that a bonefish trip to Belize, especially during spring break, would have been nice for both of us. Living in Maine, Barb was not taken with the prospect of heading even farther north to Labrador and its black flies. Given our subsequent divorce, perhaps I should have been more accommodating.

I do not consider myself a trophy hunter. I doubt any brookie angler could claim that label. In most places, a twelve-inch fish is a big fish, but Labrador undeniably represents North America's primo destination for trophy brook trout. There, brookies can grow bigger than ten pounds and reach more than twenty inches. Compared with the six-inch fish I caught in the Green, White, or Smoky Mountains, brookies in Labrador barely resemble the same species. But the Labrador giants, with blaze-orange bellies, kype jaws (lower jaws with a hook), and bronze backs, are indeed brook trout.

Trophy hunting these fish also can be dangerous. Who does not imagine catching a lifetime fish? It is easy to get caught up in the fantasy or even become obsessive after seeing the huge fish featured in lodge advertisements. When one's mind runs wild with fish fantasies, it is easy to be disappointed when things don't work out exactly as planned. "Careful," I told myself, "don't forget the Second Noble Truth: *The origin of suffering is desire.*" *Desire, in this case, was a trophy brook trout.*

As the 1950s vintage red-and-white Otter pontoon plane headed out of Wabush toward Three Rivers Lodge, where I would spend the next week, the clutter of civilization—houses, oil drums, mine waste, and prefabricated buildings—abruptly disappeared. I saw a scattering of camps near the town but then only a single empty dirt road and one endless rail line. No more than thirty minutes later, all vestiges of human habitation disappeared. Low hills interspersed by lakes, ponds, bogs, short rivers, and taiga forest stretched to the horizon. Wilderness! This was why I had come so far, more than a thousand miles northeast of Maine; this was why I had spent more money on a week-long trip than I ever had for another fishing adventure.

The hills were smooth and rounded, their bald tops exposing their foundation of gray, white, and black granite. Glaciers from the last ice age had removed any jagged features on the tops of those mounds. It appeared that ice had taken even the soil, which limited the growth of any vegetation beyond mosses, lichens, and scattered shrubs. The glaciers had scoured the uplands clean tens of thousands of years ago, and I saw more shades of green than I could ever have imagined: the dark green of spruce trees, lime green of bog plants, frosty green of moss, and olive green of deep lakes. A. J. Casson, the twentieth-century landscape painter and one of Canada's Group of Seven artists who famously painted nature scenes in neighboring Ontario, captured this northern landscape in many of his works. Specifically, the view from the plane reminded me of his *Halfway Lake* (1968), with its smooth dark hills and autumn sky.

As we flew north to the camp, the many hues of green were intermingled with the white caribou moss that grows on the otherwise bald granite and among the increasingly sparse and shorter evergreens. Paths worn in the moss by caribou migrations—line after line of caribou in search of grazing—wove through the forests and over the granite. I saw no caribou, but I could imagine a herd of hundreds, if not thousands, of woodland caribou filing by single file and etching the earth. According to guides I met later in the trip, the moss grows extremely slowly—one inch per half century—so, once made, paths remain visible for many years. Caribou adjust their migration according to the availability of moss in specific areas.

Caribou used to migrate within sight of Three Rivers Lodge from their calving grounds a few hundred miles to the north, but the camp's owner, Robin Reeves, told me they have not appeared since 2004. He speculated that the recent hot summers, coupled with expanding hordes of black flies, have led caribou to remain in more northern areas in Labrador and Québec. According to Reeves, several caribou-hunting outfitters south of his camp had gone out of business for lack of game, and as of 2022, Three Rivers no longer offers caribou hunts.

This trip was my first visit to a brookie landscape without serious human-created environmental problems (other than the effects of climate change). Once we were out of sight of Wabush, I saw no dams, mines, or clearcuts. From horizon to horizon, I saw nothing but wilderness. I imagine this area has looked the same since the glaciers melted more than ten thousand years ago.

Three Rivers Lodge sits on a small dry peninsula near some of the best brookie water in Labrador. It is close to several waterways that hold big brookies, and though they are considered rivers, they did not resemble any rivers I was used to. Because I grew up in St. Louis, I have long associated rivers with endless flows of vast amounts of water such as the Mississippi and Missouri. But these waterways in Labrador were short and rough, much like the Rapid River in Maine, and interspersed with calm runs that were the realm of big hungry northern pike. Even big brookies do not spend much time in slack water, and the scars on the sides of some trout I caught during the week attested to the presence of sharp-toothed pike. Instead, brook trout dominate within the rough-and-tumble rapids that are too swift to wade in many areas. I was rather amazed to see how segregated the river is, with brookies holding in the white water and pike, visible from the bank, lurking in the quiet areas beyond the current near shore. The two species are geographically close together yet aquatically worlds apart.

Three Rivers is a simple remote camp. Each cedar-shingled cabin has a porch that overlooks the water. There is a large communal knotty-pine dining room with a woodstove that keeps the coffee warm throughout the cool summer months. Although it was quite comfortable, it is not a luxury operation, but I was not after luxury. Homestyle cooking—chicken, caribou, meatloaf, and steak—served in big portions, all around a single giant table, made it feel more laid-back than some of the higher-end lodges I have visited. Nor was there the social stratification that separates clients and staff at high-end lodges. I was one of only four late-season guests, so we just blended in with the staff. The kitchen and dining room were essentially one big meeting hall under the gaze of mounted woodland caribou and beautiful hand-carved trophy-sized brook trout, so everyone—cooks, guides, and guests—socialized with everyone else. And after dinner, we shared good beer and good stories.

On the afternoon of my arrival, I began my Labradorean quest for big brookies on a river called Upper Eagle, one of the longer stretches of brookie waters I fished. This river was wide and deep and impossible to wade across. Even close to the shore, I had to be cognizant of my steps, because one slip and I would have plunged over my head into a black hole. This was not wading by standing on cobbles against a rushing current; it was boulder hopping on huge rocks just below the surface. Their slipperiness made a wading staff critically important.

I never use a wading staff, but I had brought along a hiking pole just in case, and I was glad I did. Because it was a hiking pole, I had to improvise and use a piece of old of rope to tie it to my waders. Another angler, Mark Paulek, accurately dubbed it the hillbilly wading staff. It was not the most photogenic arrangement, and I would never make the cover of an Orvis catalog, but it kept me standing upright in many dicey situations. An incredible insecurity took hold of me as I waded in deep water atop submerged boulders. When I took that step or leap of faith, I was never quite sure how slippery the surface would be or how stable that landing spot was. A boulder was not likely to move, but as I peered through murky moving water and searched for where to place my next step, the river felt and looked completely foreign. My steps were awkward, shaky. But I stayed upright.

The week started out with bright sunny skies. I had packed for fall, but the weather more resembled late summer. Melancholy skies had greeted me in Wabush, but that introduction to Canadian weather was short lived given the sunny skies at the lodge. While I did a great deal of blind casting, hoping for a random fish, most of the fishing on my first day was sight casting to specific big brookies. Fish either hovered near the bottom of deep pools or fed on the surface on the occasional small blue-winged olive mayflies. The lodge's guides know these short rivers so well that they are able to find fish in specific holding pools.

While this knowledge came in handy, I also wondered how many flies these fish had seen that summer. I had sight-casted to brookies before, but those fish were usually no larger than eight inches and found in small mountain streams. While it is a thrill when any targeted fish—no matter the size—takes your fly, casting to a three- or four-pound brookie is nerve-wracking, because there are so few places where you will ever see such a fish. I initially "fished tight" that day, because the sight of huge brookies was so exciting and new, and as a result I was a little off-target with my casts or slapping the water with my line. I wanted every cast to be perfect, which is impossible when I am not relaxed or feeling fluid. Bad hard casts usually mean no fish. I did not want to screw up, because I had no idea if I would catch another fish all week.

Unlike brookies I had caught previously, those fish were not particularly shy in that they did not bolt after my initial cast hit the water. That does not mean they would eat after repeated casts, but, again, they also did not necessarily flee immediately. So as the day progressed, my nerves settled, and I realized they were not spooked by my presence or my passing fly. After several drifts a four-pounder flashed its orange belly as it lunged at my prince nymph. Five minutes later, my guide, Ned Whittle, netted the biggest brookie I had ever caught—and it also happened to be my first fish of the week! The fish was shaped like a dark football, thick and round. I held it tightly while Whittle snapped a couple of photos before I held it in the current and watched it dart off. I almost couldn't believe it: the biggest and most beautiful brookie of my life was the first fish of my week. I kept repeating those facts, as if to convince myself I was not dreaming. I was charged up after catching the big Maine brook trout, but the Labrador fish had more size, more color, and fantastic girth.

Catching and releasing the first fish on any trip is always a relief because, at the very moment the fish swims away, I feel like the entire trip is a success. Of course, the initial thrill soon wanes and I want more fish, but even with the addictive hit of adrenaline that comes with each strike and fish, the first one is always the most special. At that moment, the trip already seems worth it.

Those fish were not especially shy, but they were picky, especially as the day wore on and the sun rose high in the sky. When I see a trout, the next step is to visualize it taking my fly. See fish, cast to fish, catch fish; it is supposed to be that simple. But some of those fish refused any offering. Those were the same spots and probably the same fish that anglers had targeted all summer, so by mid-August I should not have been surprised that they were getting tired of seeing countless different colorful and sometimes outrageous imitations of their food.

Anglers at Three Rivers Lodge release every brook trout they catch, so perhaps I was overanalyzing the situation and overestimating the fishes' memories. At the urging of my guide, I changed flies every five or so casts when I was targeting a specific fish. Sometimes it worked, and the fish ended up in the net, but at other times the fish ignored my flies. The most frustrating moment for an angler is when it becomes obvious that a visible fish will not eat a fly, which

means eventually walking (or wading) away from a big trout. Walking away from a trout, however, was the last resort. I did not give up easily, so I fished extremely slowly, sometimes resting the pool so as to make my fly seem new and appealing, at least in theory. This was my approach during the first couple of days: slow methodical fishing, with some nice fish as a reward.

Casting a large streamer all day in windy conditions is harder than anglers who don't fly-fish might realize. My shoulders and back ache not only because of casting but also because I always carry too much in my fly vest. I typically bring along five or six fly boxes and other supplies such as extra leader, tippet, and bug dope; hemostats for removing flies from fish or my ears; a rain jacket; and, on this trip, my large and heavy camera. Because I was changing flies so often, I actually was glad I had brought so many fly patterns; I used just about every type of nymph and streamer I owned. My fly-drying "shelf" on the outside of my vest was a tangled mess with a few dozen streamers and nymphs and various lengths of tippet that had become hopelessly knotted as the day wore on.

At the beginning of any fishing week, I cannot imagine ever being tired of fishing, and I always plan to fish from dawn to dusk. But after eight or nine hours of constant casting, I am ready for the evening meal and a few beers. And that is the beauty of staying at a lodge: fish hard all day, and come back to hot meal, cold beer, and time to share the day's stories. I am always the lone professor; most clients at lodges are in business or doctors, and all are male. But what is wonderful about fly-fishing is that, even though I might have little in common with other anglers, we quickly form a common bond around fly-fishing and the day's events. Politics, jobs, and even families take a backseat to a review of the day's conditions, tactics, and the sometimes-inflated results. It is rare to meet anybody but a really serious angler at such a remote and expensive lodge as Three Rivers, so comparing notes is also an educational experience. I have learned from other anglers about new tackle, tactics, and areas to fish.

Beyond good stories, the other anglers at Three Rivers were great people, which was fortunate because the group was small. Mark Paulek and Tony Kuznik, now lifelong fishing friends from Minnesota, kept us entertained late into the night with hilarious stories about their annual fishing adventures in western Canada with an absurd trout bum named Canada Joe. And Bo Ackerman, an ophthalmologist from Georgia, was the consummate gentleman with remarkable angling skills. The sun set late, well past 10:00 p.m., so our small group had plenty of time to reminisce about past trips, but the evenings passed all too quickly.

On a few nights, the northern lights took the sun's place. While they were not as bright as they are later in the year, it was still dreamlike to see lights appearing over the spruce-covered hills. The illuminated streams almost resembled a distant drive-in movie theater. Loons called, their calls from dark lakes haunting but peaceful, unlike my encounter with one in Maine.

Other than the loons, the area was almost quiet, day or night. Wind rustled the trees as temperatures rose during the day, but I heard few birds and no sounds other than the engine of the pontoon plane as it occasionally came and went. A black-backed woodpecker made an appearance in camp, pine siskins silently preened and ate seeds

and insects in the spruce canopy, a silent eagle soared up a river one afternoon, and spruce grouse crossed the river in front of me one day, but these were sporadic encounters. Most birds—at least the flocking songbirds—apparently had already headed south or perhaps I was too focused on fish to notice their presence.

Fishing late in the season comes with risks and rewards. River levels have dropped, so sight fishing for big brookies is possible. But, as the weather cools and fall approaches, brookies begin to pair up for the spawning season, and when fish begin to pair up, they are not much interested in eating, or at least eating flies. I am not especially enthusiastic about fishing for trout when they are in their full-on spawning mode because I don't think its ethical. But these brookies appeared to be a few weeks away from the heart of spawning season.

We also had some hot days and warm nights at the start of the week, so summer pushed back against the seasonal clock. By August, late summer in this part of the world, insect hatches are also few and far between, so dry-fly fishing is sporadic at best. I saw some blue-winged olive mayflies and even a few giant stone flies skittering across the surface of the water throughout the week, and fish occasionally rose to the surface to feed, but, according to my guide, Ned, these hatches did not compare with the number of insects on the water during June. I did manage to catch a couple of nice trout that first day on small dries, but that represented the best dry-fly fishing of the entire week. When a few bugs appear I always feel great anticipation and excitement because picky fish can become aggressive in an instant when food, and sometimes large quantities of it, is briefly available. During my week in Labrador, the few mayflies I saw were teases, so I relied on throwing big streamers and nymphs.

Days at the camp began with a big breakfast with heaping piles of thick-cut bacon and steaming plates of scrambled eggs before we headed out in pairs to various rivers around the area. We reached some destinations by boat; others required the floatplane. Flying out in a pontoon plane made the destination seem more exotic, even though some waters were not far away, just impossible to reach by boat. Flying was always dicey, though, because small planes are so dependent on clear weather. The floatplanes had to land on lakes that often were dotted with boulders just below the surface, so the pilot had to have total confidence about the watery runway. On flying trips, guides always packed a tent in case the weather dictated an extended stay on the stream. While I would have welcomed the opportunity to fish late and then early on most occasions, the black flies were usually so fierce that I was relieved to hear the whine of the engine at the end of the day.

After the first few days produced fish in the two- to three-pound range and a couple closer to five pounds, the brookies largely shut down midweek. Our guide, Ned, could not find fish in their usual holding pools. They were either on the move, looking for spawning partners, or they sensed the larger change of seasons or a shift in the daily weather, which seemed to occur in a matter of minutes. The morning of day 3, for example, was bright and clear, but by noon a thick wall of clouds, mist, rain, and then dark fog had rolled in and obliterated our pleasant Indian summer. Ascertaining why fish are at times aggressive and other times passive is impossible, but when fish are not responding to your fly, you examine and reexamine every

Delicate stone fly nymph (*Alloperla delicate*).
Drawing by Frederick Steinberg.

possibility, and the weather probably plays a big role. No matter the explanation, though, casting hour after hour to a seemingly empty river was tedious. Even among gorgeous scenery, I found it hard to enjoy the moment for fear of missing a fish. It's a sick obsession.

Pretrip expectations can also play with your mind during a slow stretch. I almost felt entitled to catch fish because of the money I had spent and time it took to get to Labrador. According to many guides with whom I have fished, this is a pretty typical feeling and something they understandably resent. In hindsight, the belief is silly—dare I say stupid?—but in the moment, when my shoulders ache and boredom sets in after several hours of casting without even a strike, I turn over all kinds of scenarios in my mind, some rational and some irrational: Am I a bad angler? Does my guide know what he is doing? Do I not have the right flies? Am I making poor presentations? And, silliest of all, are there *any* fish in this river? No one else at the camp was catching much either, so at the end of a long day, I stopped playing the mental blame game, which made for a better night's sleep.

One of the many nice things about Labrador and similar northern latitudes is that, when the trout slow down or disappear altogether, northern pike are always willing to bite. Pike are funny creatures because they sit in the shallow slack water and see you as easily as you see them, yet they seemingly are happy to crush your fly. Many hardcore trouters hate pike because they eat trout. Some even kill them by throwing them up on the bank, but I have come to appreciate pike, because they can rescue an otherwise slow day. Besides, where they are native, they have just as much right to exist as a trout and play a significant ecological role in a stream. We have a habit of persecuting predators, whether they are wolves, bears, or toothy fish. Many of us cannot seem to recognize the larger role these animals play in the environment and instead focus on how they might affect our favorite game. If they are nonnative, then fine—remove them. But I will not kill a native fish just because it preys upon a more desirable game fish.

During one especially slow day, Bo Ackerman, my assigned fishing partner, decided to tie on some huge gaudy streamers and mouse patterns and give pike a try. Bo was a fantastic caster, able to throw those big flies long distances even in the face of strong winds. I have never considered myself a stylistic caster, and sometimes it is fun just to sit back and watch someone really throw it. Just like Norman Maclean would watch and admire the casting artistry of his younger brother, Paul, in *A River Runs through It and*

Other Stories (1976). Soon Bo had hooked and released several pike, including a forty-four-incher. Its tail, almost eel-like, hung over the side of the net. Because pike seemingly strike at anything, it seems they are either dumb or just plain ornery. I vote for latter. After seeing that monster, I needed to get in on the action. After a couple of casts among the reeds, the water erupted just a few yards from the end of my fly rod, as a nice pike slashed my streamer. Ten minutes later, a thirty-five-inch pike was in my net. Ned immobilized the pike by pressing on its eyes while removing the fly. When I asked if this hurts the fish, he responded in his dry Newfoundland manner, "They sure as hell don't like it." He had no interest in getting near those rows of razor-sharp teeth.

The river we fished that day also held whitefish, lake trout, and landlocked Atlantic salmon (the last migrate between lakes), so who was on the end of the line was always a bit of mystery. Like other rivers in the area, this one was short; unlike the others, it was mostly shallow and wide, with a cobble bottom. It resembled the arctic rivers found in many parts of Alaska and felt like one, with a cold steady rain and howling wind. I was fishing an isolated deep hole at the end of a long shallow riffle. It was easy fishing because I could stand in shallow water but have access to a deep pool I knew held large fish.

This was a perfect spot, as a lot of the river's water fed into this one area, carving out a large dark run that just reeked of big fish. The water was too swift for pike, so I was confident big trout sat in that pool, picking off meals flushed down by the constricted current. Sure enough, after a few casts, I hooked a beautiful, three-pound brookie on a streamer. It probably was the prettiest fish of the week. It definitely had its mid-August spawning colors on, its belly ranging from bright orange to dark yellow like the shades of a brilliant summer sunset. Its fins were a burnt rust color, which complemented its dark brown back. In this spruce green and granite gray landscape, the brook trout provided most of the color, especially on a cold wet day.

Although the day had been slow in terms of brookies, that fish was handsome enough to make the hours of fish-free casts worth it. I did not rush it. I knew the fish was not especially big for Labrador, but I did not want to blow my opportunity to catch a nice fish on a slow day. Even large flies for trout have relatively small hooks, and my flies had the barbs pinched down to reduce the possibility of injuring the fish, so forcing a nice fish to my net prematurely often resulted in no fish at all. I admired the fish for a few moments and sent it back to its dark depths.

Several casts later, another fish ran with my olive bunny streamer. I knew immediately this was not a three-pound fish, because I was quickly losing my fly line. My reel screamed, as the fish ran downriver, and I was soon deep into my backing. Ned yelled to follow the fish downstream before it spooled my line. I was standing almost dumb-struck as the fish fled and my reel screamed. Luckily, I was able to follow the shallow riffle that skirted the deep hole down the river for another hundred yards, so the wading was relatively easy. As I tightened my drag, I knew this was not a northern pike. If this was a brookie, it was in the ten-pound class. As I slowed the fish and reeled in line, I caught my first glimpse of his tail. It was indeed a trout—but a laker, not a brookie.

While I admit to some disappointment, I had never caught a lake trout on a fly, let alone a fifteen-pound fish

on a six-weight rod. It was a great fight, so I had no reason to complain. I was there for brookies, but a great fish is a great fish. The laker finally slowed, and I was able to get much of my line back on the reel. Ned was able to get below the fish to net it. I had hooked it cleanly in the lip, so even after a hard fight, the fish swam easily away once I snapped a picture. I greedily waded back up to the hole where I had hooked that fish, hoping that a giant brookie might still be lurking. Ned laughed at my gluttony. After the ruckus caused by my laker, the pool was quiet. No matter—I needed to rest as well.

I had to hold on to the memories of those two nice fish, because the next two days were brookie free. For a solid day, I fished the Five Rapids area, said to be one of the best sections of river in the area, but came away with zero trout. Ned claimed the fish were not in the spots they normally would be. My fishing partner for the day, Mark Paulek, was able to land a nice brookie from a small plunge pool, but that was the only one between the two of us. We fished hard, hitting every trouty-looking run and pool. The lodge's guides knew exactly where to look, but nothing much was happening. A few fish followed my streamer but not aggressively. I enlisted the help of my childhood Catholicism, reciting the Hail Mary over and over, to no avail. Even the Blessed Mother had abandoned me. As blasphemous as that sounds, I was desperate.

Fishless days are most frustrating, because you can feel your trip slipping away. Days and even casts are precious commodities to be savored during a remote trip. That's what I try to tell myself after seemingly hundreds of casts without a fish. Maintaining focus during those days is hard. At one point in the early afternoon, I saw a nice brookie rising to a tiny unidentifiable insect. I never really saw the insect at all. Whatever the fish was eating, it was small and probably just emerging, but I never had a good look at whatever was drawing that fish to the surface. The fish was big, probably four pounds. It kept rising within an area the size of a bathtub, only fifteen or so feet below my boulder platform. It moved a few feet to the right and left and up and down in the current but stayed within the same small area the entire time. I literally cast to that single fish for more than two and a half hours. Two and a half hours! And I literally tied on every dry-fly pattern I carried with me as well as many streamers. The fish rose to my fly many times, with a tiny Adams drawing the most attention. But it would not strike; instead, it rose just close enough to see what I was offering and then dropped back down again. I saw its golden belly as it rose and dove without my fly, time and time again. It sort of sniffed the offering and said, "No way." I tried everything from large stone-fly patterns to tiny size-24 midges; the latter forced me to tie on a lighter tippet. Ned thought I was crazy. Shaking his head, he yelled out, "You are stubborn!" Mark, meanwhile, fished all around me and up and down the river. Above the din of the rapids, he occasionally would call out, "Any luck?" But based on my expression, he did not wait long for my answer before continuing his own frustrating hunt for stubborn fish.

At the time, I wrote off the trout's refusal to cooperate as just a challenge, not something to stress about, because stress results in bad casts. But on a fishless day, watching a big brookie reject cast after cast is simply frustrating. I have heard and read stories about anglers who carry minimal gear. For example, in his classic story Norman Maclean says Paul, his brother, carried only a few basic fly patterns. While I like that simple straightforward approach, on that day I wished I had even more options. But I had offered about forty fly patterns to that one fish, so I am not sure that more choices would have mattered.

Black skies and a late-season thunderstorm rolling in over the nearby hills finally drove us away from that fish and back to camp. I really wanted to keep trying because walking away from a feeding fish on a day when I had caught zilch is tough. But the weather gods probably did me a favor. Had I continued to beat that pool for much longer, I may have temporarily lost my mind.

During such fishless times, I try desperately not to become obsessed by the lack of success. According to *The Tibetan Book of the Dead* (1326–86), your last thoughts in this life will determine the form you take in the next one. I have thought of that belief when frustration sets in during long tedious days. If it were my last day in this form, in what lowly state would I return? A remora or a chub? Does the lack of fish really warrant anger? Alas, I always think of that question after the fact. Peter Matthiessen wrote in *The Snow Leopard* (1978): "When one pays attention to the present, there is great pleasure in awareness of small things." While I recognize the wisdom in this simple statement, I often disappoint myself because taking pleasure in small things is an impossible task when I fish indignantly and the fish are not cooperating. In hindsight, my attitude makes me even angrier because I realize I probably missed so many small things—new birds, insects, or simple wildflowers—all because I wanted a fish too much.

When fish are aggressive, ignoring unpleasant distractions such as biting insects is easy. When the fishing is slow, though, those bites seem worse than they are. By midweek, when fishing ground to a halt, the black flies seemed hungrier and more populous. I know my imagination was running wild, perhaps literally carried away by flies, but their attacks did seem to intensify. I had never ever experienced bugs like I did that week, and I had fished for years in Maine, which is notorious for its own plagues of black flies. Those bugs in Labrador found their way into my ears, eyes, nostrils, shirt, up my sleeves, and into my lungs. My hat was literally covered with black flies trying somehow to dig through to my scalp. They attacked any area, even the smallest imaginable, if it wasn't covered with Deet. My eyelids became bruised and bloody from their bites. My lips were swollen. Small streaks of blood ran and then dried on my face and in my beard, crusty reminders of the day's pain. Given the flies, could anyone really call this home water? The guides knew and loved the area, but even they were seasonal and loathed the black flies.

My face was so chewed up that Mark nicknamed me "Bloody Beard Mike." I looked and felt like a street fighter in waders. I refused to wear the head net that he offered because I had trouble wading in it. Boulder hopping is difficult enough when all my senses are tuned in; I did not need the impairment of both my sight and balance. So, in addition to using 100 percent Deet, I worked to ignore them like a master of mindful meditation might suggest. I would try as hard as possible to accept the flies, accept their presence, and accept their bites so I could focus on the task at hand.

I literally had hundreds, if not thousands, of bugs swarming around me, crawling into the corners of my eyes, creating painful, dizzying, temporary blindness. But on a few occasions, I was able to enter a Zen-like meditative state and tune them out. I was able to focus my mind and concentration on the river and fishing and separate them from the bugs. Of course, I usually could enter this state while also catching fish, but sometimes I psychologically beat those insects for long stretches. I also felt that the bugs and their wrath were part of the price I had to pay to fish in Labrador. If I could not handle the bugs, I did not deserve to catch fish. All pilgrimages come with risks and rewards along the way.

Much of Labrador is covered with wet boggy spruce forests, the perfect breeding ground for black flies and mosquitoes. I almost longed for the mosquito-control units from my teaching days at Louisiana State University when, late at night, a truck would drive through the streets of Baton Rouge and spray a probably toxic brew that kept the subtropical bugs at bay. As the hum of its blower became louder, Barb and I would spring from bed to close the windows before the mysterious fog settled in the house. Where was that truck when I needed it? As the sun set and the temperature dropped, the flies quieted, giving everyone a few hours of peace to drink beers and smoke cigars. Without the evening break from the flies, we would have gone mad.

Given the slow days at midweek, the end of my trip came much too quickly. I did not want to leave Labrador in a frustrated fishing mood. Don't get me wrong: I had caught some nice fish the first few days. But I had to break my midweek slump. Bo and I were fortunate to be assigned a river that had hardly been touched during the summer, and we had a new guide, Cliff Randell, who had fished it once the previous year. It sounded like this was a fallback river, a sure thing; its brookies had not seen hundreds or thousands of wooly buggers. Fishing near the bottom of the river, where it emptied into a large lake, Bo right away caught two big fish more than five pounds each, the biggest brookies I had ever seen. I had not tied on a fly before he landed his first big trout. Those early fish made his day. I was more than a little envious, but at the same time I also was happy for his success.

I worked my way upstream, about fifty yards past some marginal-looking riffles, and caught several nice two-pound fish, all from a small holding spot behind a boulder. I knew at that moment it was going to be a great day, the kind I had dreamed of when I booked my trip. They were not huge fish, but they were nice and aggressive. After two nearly fishless days, my bad luck had broken. Near the same run, I missed an even larger fish across the stream. I resisted the temptation to throw my fly again immediately and waited several minutes. On the second cast, the fish rose again and flashed at my streamer, more tentatively this time, refusing to take it. I moved on, making a mental map of its location, hoping to reacquaint myself with that fish on my way back downstream at the end of the day. Catching a challenging fish, one that is big and wily, is a great satisfaction, in part from making the proper cast and presentation. But it is deeper or more primordial than that. I get satisfaction, or even a rush, from literally being connected to a fish, fighting something that is alive. When I feel the strength of a big fish hooked to a tiny fly on the end of my line, I feel more alive and more connected to nature.

Although that stream was smaller than any I had fished earlier in the week—I was able to wade across it tentatively in a few places—it was full of hungry solid fish. Again, these were virgin fish, and unlike some of the bigger pieces of water I had fished earlier, this river had not been fished all summer. Many fish were in the two-pound range, not trophies; before my trip to Labrador, that would have been a lifetime brookie. Bo and I laughed about that after we had grumped about smallish fish during lunch (peanut butter with a generous side of black flies). As we fished up the river, jumping past one another to the next pool, we often caught fish on seemingly every cast. Amazing. Like the other rivers we fished that week, our guide knew this one well. He knew exactly which pools and parts of the river held the biggest fish. Even in the slower stretches, places I normally would expect to find only pike, I caught brookies. Amazing. I was out of breath with excitement at times because there were fish to be had on every cast.

It was shaping up to be by far the best day of the week and perhaps the best single angling day of my life. By late afternoon, we had each caught sixty, then seventy, maybe eighty fish. It was the sort of fishing that happens infrequently, to put it mildly. This stream had no worn and muddy path along its bank. I still had not caught a fish that weighed more than five pounds, the real trophy brookie upon which Labrador's reputation has been built, but I had caught and released several three- to four-pound fish. And I had lost a couple of even bigger fish in the swift currents of the larger rivers at the beginning of the week. Even as that trophy fish still eluded me, I told myself I did not really crave it anymore, not after a seventy-fish day. I am not that crazy or that spoiled. But as I worked my way back downriver, catching fewer fish, in the back of my mind I thought I might leave Labrador without the memory of a trophy. Maybe I would again fish the pool with the big brookie that earlier in the day had flashed at my streamer.

I was on the other side of the river at that point, so I had to crawl through streamside stunted willows and make a sort of acrobatic cast to reach the area where the fish had risen earlier in the day. I caught it on the first cast and drift on a big muddler minnow streamer. The fish was beautiful—dark bronze, thick, and four pounds. I was satisfied I had approached that fish's run methodically, made an accurate cast, and landed the biggest fish of my day. That is how fly-fishing is supposed to play out when you do things correctly: you do your part, the fish cooperates, and then you release the fish and watch it swim away.

Because I was near the bottom of the river at the end of the day, only one pool or run was left to fish, and that was the area where Bo had landed his big fish at the beginning of the day. It was still his spot, because he had fished it earlier. But he is a great sportsman and insisted I fish the area for one last shot at a really big fish. He walked down past the end of the river and fished the lake for pike, tying on a fly so large that it resembled a road-kill squirrel. The floatplane would be arriving within the hour to take us back to camp. My week was rapidly winding down. The last hour of fishing on any remote trip is an unsettling experience. Yes, I took great satisfaction in my angling success during the week and the new friends I had made, but knowing that my last cast was more than likely my last cast *ever* in that particular landscape was deflating. I never want to admit I might not visit a specific river or camp again, Labrador or elsewhere, but, given the cost and the plethora of fishing options around the globe, I knew it was unlikely I would ever fish that area again, which added to my suddenly gloomy mood.

I started casting a large olive bunny-hair zonker, my mainstay at that point, around a huge boulder right at the bottom of the river where it entered a long slack lake. I had no luck for about the first dozen or so casts, but I had time, so I kept throwing the fly. I cast the streamer right up against the boulder, literally bouncing it off the side of the rock, when a huge brookie breached the water and apparently took it. I say apparently because I thought the fish lunged at the streamer without taking it, given that the fly had just hit the water. But I reflexively set the hook, muscle memory at that point in the day, and much to my surprise, the fish was on.

I was fortunate when the brookie ran downstream into the slack water at the mouth of the lake. I had to fight this fish cautiously and give him as much line as he wanted. I was not going to force this fish to the net. He ran up and down the slack water, taking line and then giving it back. My only challenge was to keep him off the bottom, where he might dislodge the fly or wedge himself under a rock. He never broke the surface, but I knew it was a brookie based on his initial lunge at my streamer. And my heart pounded with excitement, because this was clearly my biggest brookie of the week. I had a stout leader on, with a big fly, so I felt confident he was not going anywhere, but I knew this was my true trophy fish—if I played it right—and my last chance. I did not want to blow it.

Finally, after about fifteen minutes, I had most of my line in, and Cliff, our guide netted, the fish. My wrists quivered from the tension, but I could exhale now. This brookie was an incredible twenty-two inches, eight pounds, and as wide and round as a football, or so it seemed. I spent a few minutes reviving the fish in the current, and with a flick of a tail he was gone. Reviving a big fish is part ceremony. I never would have forgiven myself if my catch had died in the net. I exhaled deeply, truly grateful for my luck and for Bo's generosity in giving me that last shot. I then whooped and hollered. I shared high fives with Bo and Cliff.

I could not have imagined a better ending to my week and my quest. My last fish of the week was the biggest fish. I have thought about that scene over and over ever since. And the irony or luck. Perhaps my fishing karma was rewarded because I had overcome my desire for a trophy. I was simply taking in the last hour before we called it quits for the day and trip. I had not expected a huge fish. I was more than happy with the week overall and especially that specific day. Whatever frustration I had experienced was largely the result of my ridiculous expectations that every day would yield more fish and another trophy bigger than the previous day's.

After releasing my trophy, I still had time to fish, but I thought better of it. I broke down my rod, sat on a nearby boulder, swatted black flies, breathed deeply in the late afternoon sun as the floatplane approached in the distance, and savored the moment. Indeed, I had seen the snow leopard, and it was wonderful. It was time to go home.

A Great Smoky Mountain National
Park brook trout in fall colors.
Photograph by Robin Boggs.

Sources

The Eastern Brook Trout as Icon

Jim Babb, *River Music: A Fly Fisher's Four Seasons* (New York: Lyons Press, 2005).

Gary Berti, *Environmental Protection*, May 1, 2006, https://eponline.com/articles/2006/05/01/study-brook-trout-imperiled-throughout-eastern-united-states.aspx.

John Burroughs, "Speckled Trout," in *In the Catskills: Selections from the Writings of John Burroughs* (Boston: Houghton Mifflin, 1910).

Chris Camuto, *A Fly Fisherman's Blue Ridge* (Athens: University of Georgia Press, 2001).

John Gierach, *Fly Fishing Small Streams* (Mechanicsburg, Pa.: Stackpole Books, 1989).

———, *Trout Bum* (New York: Simon and Schuster, 1988).

Nick Karas, *Brook Trout* (Guilford, Conn.: Lyons Press, 2002).

Nick Lyons, *A Flyfisher's World*, illustrations by Mary Lyons (New York: Atlantic Monthly Press, 1998).

———, *Brook Trout* (Guilford, Conn.: Lyons Press, 2002).

———, *Confessions of a Fly Fishing Addict* (New York: Atlantic Monthly Press, 1999).

———, ed., *In Praise of Wild Trout* (Guilford, Conn.: Lyons Press, 1998).

Norman Maclean, *A River Runs through It and Other Stories* (Chicago: University of Chicago Press, 1976).

John McDonald, *The Complete Fly Fisherman: The Notes and Letters of Theodore Gordon* (New York: Lyons Press, 1989).

Craig Nova, *Brook Trout and the Writing Life: The Intermingling of Fishing and Writing in a Novelist's Life* (Hillsborough, N.C.: Eno, 2011).

James Prosek, *Early Love and Brook Trout* (Guilford, Conn.: Lyons Press, 2000).

Henry David Thoreau, *Walden, or, Life in the Woods* (Boston: Ticknor and Fields, 1854).

W. D. Wetherell, *Vermont River: The Classic Portrait of a Man and His River* (Guilford, Conn.: Lyons Press, 1993).

Home Waters and Symbols

Robert J. Behnke, *About Trout: The Best of Robert J. Behnke from* Trout Magazine (Guilford, Conn.: Lyons Press, 2007).

———, *Trout and Salmon of North America* (New York: Free Press, 2002).

Chris Camuto, *A Fly Fisherman's Blue Ridge* (Athens: University of Georgia Press, 2001).

Ernest Hemingway, *Islands in the Stream* (New York: Charles
    Scribner's Sons, 1970).
Mark Hudy, Teresa M. Thieling, Nathaniel Gillespie, and
    Eric P. Smith, "Distribution, Status, and Land-Use
    Characteristics of Subwatersheds within the Native Range
    of Brook Trout in the Eastern United States," *North
    American Journal of Fisheries Management* 28, no. 4
    (January 2008): 1069–85.
Jack Kerouac, *The Dharma Bums* (New York: Viking Press,
    1958).
Peter Matthiessen, *Nine-Headed Dragon River: Zen Journals
    1969–1982* (New York: Shambhala, 2014).
———, *The Snow Leopard* (New York: Viking Press, 1978).
Thomas McGuane, *The Longest Silence: A Life in Fishing* (New
    York: Vintage, 2014).
Herman Melville, *Moby-Dick; or, The Whale* (New York:
    Harper, 1851).
William Powell, "Remember Times Beach: The Dioxin Disaster
    30 Years Later," *St. Louis Magazine*, December 3, 2012, 1–4,
    www.stlmag.com/Remember-Times-Beach-The-Dioxin
    -Disaster-30-Years-Later/.
Mark Sosin and Lefty Kreh, *Fishing the Flats* (New York:
    Lyons Press, 1983).
Yi-fu Tuan, *Topophilia: A Study of Environmental Perceptions,
    Attitudes, and Values* (Englewood Cliffs, N.J.: Prentice-Hall,
    1974).
W. D. Wetherell, *Vermont River: The Classic Portrait of a Man
    and His River* (Guilford, Conn.: Lyons Press, 1993).
Gordon G. Whitney, *From Coastal Wilderness to Fruited Plain:
    A History of Environmental Change in Temperate North
    America from 1500 to the Present* (Cambridge: Cambridge
    University Press, 1996).

Chapter 1. Georgia

Donald E. Davis, *Where There Are Mountains: An
    Environmental History of the Southern Appalachians*
    (Athens: University of Georgia Press, 2011).
Gary L. Larson and Stephen E. Moore, "Encroachment of
    Exotic Rainbow Trout into Stream Populations of Native
    Brook Trout in the Southern Appalachian Mountains,"
    Transactions of the American Fisheries Society 114, no. 2
    (February 1985): 195–203.
Todd C. McDonnell, Matthew R. Sloat, Timothy J. Sullivan,
    C. Andrew Dolloff, Paul Francis Hessburg, Nicholas A.
    Povak, William A. Jackson, and Charles Sams, "Downstream
    Warming and Headwater Acidity May Diminish Coldwater
    Habitat in Southern Appalachian Mountain Streams," *PLOS
    One* 10, no. 8 (September 2015): e0134757. 1–23.
Donald J. Meisner, "Effect of Climatic Warming on the
    Southern Margins of the Native Range of Brook Trout,
    Salvelinus fontinalis," *Canadian Journal of Fisheries and
    Aquatic Sciences* 47, no. 6 (June 1990): 1065–70.
Katharine A. Owers, Brett Albanese, and Thomas Litts, "Using
    Aerial Photography to Estimate Riparian Zone Impacts
    in a Rapidly Developing River Corridor," *Environmental
    Management* 49, no. 3 (November 2011): 543–52.
Trout Unlimited for the Eastern Brook Trout Joint Venture,
    Status and Threats (Arlington, Va., 2006), www
    .easternbrooktrout.org.

Chapter 2. South Carolina

Susan Power Bratton and Albert J. Meier, "The Recent
    Vegetation Disturbance History of the Chattooga River
    Watershed," *Castanea* 63, no. 3 (September 1998): 372–81.
Hazel R. Delcourt and Paul A. Delcourt, "Pre-Columbian
    Native American Use of Fire on Southern Appalachian

Landscapes," *Conservation Biology* 11, no. 4 (August 1997): 1010–14.

James Dickey, *Deliverance: A Novel* (Boston: Houghton Mifflin, 1970).

Michael D. Enk, "Instream: Overhead Bank Cover and Trout Abundance in Two Michigan Streams," unpublished master's thesis, Michigan State University, East Lansing, 1977.

Donald J. Meisner, "Effect of Climatic Warming on the Southern Margins of the Native Range of Brook Trout, *Salvelinus fontinalis*," *Canadian Journal of Fisheries and Aquatic Sciences* 47, no. 6 (June 1990): 1065–70.

Trout Unlimited for the Eastern Brook Trout Joint Venture, *Status and Threats* (Arlington, Va.: 2006), www.easternbrooktrout.org.

Chapter 3. Tennessee

John Preston Arthur, *Western North Carolina History from 1790 to 1913* (Raleigh, N.C.: Edwards & Broughton, 1914).

Daniel M. Evans, Michael Aust, C. Andrew Dolloff, Ben S. Templeton, and John A. Peterson, "Eastern Hemlock Decline in Riparian Areas from Maine to Alabama," *Northern Journal of Applied Forestry* 28, no. 2 (June 2011): 97–104.

Samuel J. Hunnicutt, *Twenty Years of Hunting and Fishing in the Great Smoky Mountains* (Maryville, Tenn.: Byron's, 1951).

"Recreation Groups Express Disappointment at FS [Forest Service] Implementing Tellico Emergency Closure," UTVGuide.net, April 1, 2009, https://www.utvguide.net/recreation-groups-express-disappointment-at-fs-implementing-tellico-emergency-closure/.

Douglas M. Thompson, *Quest for the Golden Trout: Environmental Loss and America's Iconic Fish* (Lebanon, N.H.: University Press of New England, 2013).

Southern Environmental Law Center, "Federal Court Upholds Closure of the Tellico ORV Area to Protect Brook Trout,"

September 19, 2012, https://www.southernenvironment.org/press-release/federal-court-upholds-closure-of-the-tellico-orv-area-to-protect-brook-trou/.

Chapter 4. North Carolina

John Preston Arthur, *Western North Carolina History from 1790 to 1913* (Raleigh, N.C.: Edwards & Broughton, 1914).

Patricia A. Flebbe, "A Regional View of the Margin: Salmonid Abundance and Distribution in the Southern Appalachian Mountains of North Carolina and Virginia," *Transactions of the American Fisheries Society* 123, no. 4 (June 1994): 657–67.

Patricia A. Flebbe and C. Andrew Dolloff, "Trout Use of Woody Debris and Habitat in Appalachian Wilderness Streams of North Carolina," *North American Journal of Fisheries Management* 15, no. 3 (August 1995): 579–90.

Keith W. Gibbs, Matt A. Kulp, Steve E. Moore, and S. Bradford Cook, "Effects of Antimycin and Its Neutralizing Agent, Potassium Permanganate, on Aquatic Macroinvertebrates in a Great Smoky Mountains National Park Watershed," *North American Journal of Fisheries Management* 35, no. 6 (November 2015): 1239–51.

Ernest Hemingway, *Big Two-Hearted River* (New York: Boni and Liveright, 1925).

Gary F. McCracken, Charles R. Parker, and Stanley Z. Guffey, "Genetic Differentiation and Hybridization between Stocked Hatchery and Native Brook Trout in Great Smoky Mountains National Park," *Transactions of the American Fisheries Society* 122, no. 4 (June 1993): 533–42.

North Carolina Wildlife Resources Commission, *Brook Trout Management in North Carolina: Supplement to a Plan for Management of North Carolina's Trout Resources* (Raleigh: North Carolina Wildlife Resources Commission, 2003).

Trout Unlimited for Eastern Brook Trout Joint Venture, *Status and Threats* (Arlington, Va., 2006), www.easternbrooktrout.org.

Chapter 5. Virginia

Edward Abbey, *Desert Solitaire* (New York: McGraw-Hill, 1968).

Chris Camuto, *A Fly Fisherman's Blue Ridge* (Athens: University of Georgia Press, 2001).

John G. Eaton and Richard M. Scheller, "Effects of Climate Warming on Fish Thermal Habitat in Streams of the United States," *Limnology and Oceanography* 41, no. 5 (June 1996): 1109–15.

Nicholas A. Fisichelli, Scott R. Abella, Matthew Peters, and Frank J. Krist Jr., "Climate, Trees, Pests, and Weeds: Change, Uncertainty, and Biotic Stressors in Eastern U.S. National Park Forests," *Forest Ecology and Management* 327 (September 2014): 31–39.

Herbert Hoover, *Fishing for Fun—And to Wash Your Soul* (New York: Random House, 1963).

Dennis D. Lynch and Nancy B. Dise, *Sensitivity of Stream Basins in Shenandoah National Park to Acid Deposition*, Report no. 85-4115 (Washington, D.C.: U.S. Geological Survey, 1985).

Stephen E. MacAvoy and Arthur J. Bulger, "Survival of Brook Trout (*Salvelinus fontinalis*) Embryos and Fry in Streams of Different Acid Sensitivity in Shenandoah National Park, U.S.A.," *Water, Air, and Soil Pollution* 85, no. 2 (December 1995): 445–50.

John Ross, *Rivers of Restoration*: *Trout Unlimited's First 50 Years of Conservation* (New York: Skyhorse, 2008).

——— , *Trout Unlimited's Guide to America's 100 Best Trout Streams* (New York: Lyons Press, 1999).

Izaak Walton, *The Compleat Angler, or the Contemplative Man's Recreation, Being a Discourse of Fish and Fishing* (London: Richard Marriott, 1653).

James R. Webb, Bernard J. Cosby, Frank A. Deviney, James N. Galloway, Suzanne W. Maben, and Arthur J. Bulger, "Are Brook Trout Streams in Western Virginia and Shenandoah National Park Recovering from Acidification?" *Environmental Science & Technology* 38, no. 15 (September 2004): 4091–96.

Chapter 6. West Virginia

James T. Anderson, Ryan L. Ward, Todd J. Petty, J. Steven Kite, and Michael P. Strager, "Culvert Effects on Stream and Stream-side Salamander Habitats," *International Journal of Environmental Science and Development* 5, no. 3 (January 2014): 274–81.

Arthur J. Bulger Jr., "Blood, Poison and Death: Effects of Acid Deposition on Fish," in J. C. White, ed., *Acid Rain: Are the Problems Solved?* (Bethesda, Md.: American Fisheries Society, 2003), 59–64.

Robert F. Carline and Brian J. McCullough, "Effects of Floods on Brook Trout Populations in the Monongahela National Forest, West Virginia," *Transactions of the American Fisheries Society* 132, no. 5 (August 2003): 1014–20.

John R. Gibson, Richard L. Haedrich, and C. Michael Wernerheim, "Loss of Fish Habitat as a Consequence of Inappropriately Constructed Stream Crossings," *Fisheries* 30, no. 1 (January 2005): 10–17.

Mark Hudy, Daniel M. Downey, and Darrell W. Bowman, "Successful Restoration of an Acidified Native Brook Trout Stream through Mitigation with Limestone Sand," *North American Journal of Fisheries Management* 20, no. 2 (April 2000): 453–66.

Joseph Love, "Reintroducing Elk to the Great Smoky Mountains," *American Forests* 119, no. 2 (2013): 32–37.

Todd J. Petty, Jennifer Barker Fulton, Michael P. Strager, George T. Merovich Jr., James M. Stiles, and Paul F. Ziemkiewicz, "Landscape Indicators and Thresholds of Stream Ecological Impairment in an Intensively Mined Appalachian Watershed," *Journal of the North American Benthological Society* 29, no. 4 (December 2010): 1292–309.

Todd J. Petty, Jeff Lee Hansbarger, Brock M. Huntsman, and Patricia M. Mazik, "Brook Trout Movement in Response to Temperature, Flow, and Thermal Refugia within a Complex Appalachian Riverscape," *Transactions of the American Fisheries Society* 141, no. 4 (July 2012): 1060–73.

Chapter 7. Maryland

Chester R. Arnold Jr. and C. James Gibbons, "Impervious Surface Coverage: The Emergence of a Key Environmental Indicator," *Journal of the American planning Association* 62, no. 2 (June 1996): 243–58.

John G. Eaton and Robert M. Scheller, "Effects of Climate Warming on Fish Thermal Habitat in Streams of the United States," *Limnology and Oceanography* 41, no. 5 (July 1996): 1109–15.

Alan A. Heft, ed., *Maryland Brook Trout Fisheries Management Plan* (Annapolis: Maryland Department of Natural Resources, Fisheries Service, Inland Fisheries Management Division, 2006).

David C. Kazyak, Robert H. Hilderbrand, and Matthew T. Sell, "Growth Variation in a Mid-Atlantic Brook Trout Population," *Proceedings of the Wild Trout XI Symposium*, West Yellowstone, Montana, September 22–25, 2014, 350–56.

Maryland Department of Natural Resources, *Maryland Wildlife Diversity Conservation Plan* (Annapolis: Maryland Department of Natural Resources, 2005.

Albert K. Smith and Dann Sklarew, "A Stream Suitability Index for Brook Trout (*Savelinus fontinalis*) in the Mid-Atlantic United States of America," *Ecological Indicators* 23 (December 2012): 242–49.

Scott A. Stranko, Robert H. Hilderbrand, Raymond P. Morgan, Mark W. Staley, Andrew J. Becker, Ann Roseberry-Lincoln, Elgin S. Perry, and Paul T. Jacobson, "Brook Trout Declines with Land Cover and Temperature Changes in Maryland," *North American Journal of Fisheries Management* 28, no. 4 (July 2008): 1223–32.

George F. Thompson, ed., *Landscape in America* (Austin: University of Texas Press, 1995), 2.

Chapter 8. Pennsylvania

Ray Bergman, *Trout*, 3rd ed. (New York: Derrydale Press, 2000).

Ad Crable, "Pennsylvania Takes 'Historic' Action to Fund Clean Water Projects—And Help the Chesapeake Bay," *Bay Journal*, July 11, 2022, https://www.bayjournal.com/news/policy/pennsylvania-takes-historic-action-to-fund-clean-water-projects-and-help-the-chesapeake-bay/article_88dc1578-0118-11ed-91e7-9f50d6c12599.html.

William Flick, "Brook Trout (*Salvelinus fontinalis*)," in Judith Stohlz and Judith Schnell, eds., *The Wildlife Series: Trout* (Harrisburg, Pa.: Stackpole Books, 1991), 196–207.

Charles Fox, *Rising Trout* (New York: Dutton Books, 1978).

Josh Fox, dir., *Gasland*, independent film (June 2010).

Patrick F. Kocovsky and Robert F. Carline, "Stream pH as an Abiotic Gradient Influencing Distributions of Trout in Pennsylvania Streams," *Transactions of the American Fisheries Society* 134, no. 5 (August 2005): 1299–312.

Abrahm Lustgarten, "New Study: Fluids from Marcellus Shale Likely Seeping into PA Drinking Water," *ProPublica*, July 9, 2012, www.propublica.org/article/new-study-fluids-from-marcellus-shale-likely-seeping-into-pa-drinking-water.

Peter Matthiessen, *The Snow Leopard* (New York: Viking, 1978).

Trout Unlimited, "Catawissa Creek Watershed Restoration Plan Update Addressing the TMDL," March 20, 2007, https://files.dep.state.pa.us/Water/BWEW/Watershed%20Management/lib/watershedmgmt/nonpoint_source/implementation/catawissa_creek_plan.pdf.

Ian Urbina, "Regulation Lax as Gas Wells' Tainted Water Hits Rivers," *New York Times*, February 27, 2011, A1, www.nytimes.com/2011/02/27/us/27gas.html.

Robert Weber, Russell T. Greene, John Arway, R. Scott Carney, and Leroy Young, *History of the Management of Trout Fisheries in Pennsylvania* (Harrisburg: Pennsylvania Fish and Boat Commission, 2010).

Maya Weltman-Fahs and Jason M. Taylor, "Hydraulic Fracturing and Brook Trout Habitat in the Marcellus Shale Region: Potential Impacts and Research Needs," *Fisheries* 38, no. 1 (February 2013): 4–15.

Chapter 9. New Jersey

Patricia Hamilton, "Genetic Diversity of Wild Brook Trout (*Salvelinus fontinalis*) Populations in New Jersey: Conservation and Management Implications," unpublished master's thesis, East Stroudsburg University, East Stroudsburg, Pa., 2007.

David L. Perkins, Charles C. Krueger, and Bernie May, "Heritage Brook Trout in Northeastern U.S.A: Genetic Variability within and among Populations," *Transactions of the American Fisheries Society* 122, no. 4 (June 1993): 515–32.

State of New Jersey, *Management of Wild Trout—State of New Jersey* (Trenton: 2005).

Chapter 10. New York

Barry P. Baldigo, Gregory Lawrence, and Howard Simonin, "Persistent Mortality of Brook Trout in Episodically Acidified Streams of the Southwestern Adirondack Mountains, New York," *Transactions of the American Fisheries Society* 136, no. 1 (January 2007): 121–34.

Leo Demong, "The Use of Rotenone to Restore Native Brook Trout in the Adirondack Mountains of New York—An Overview," address at the 130th Annual Meeting of the American Fisheries Society, St. Louis, Mo., 2000.

Paul Schneider, *The Adirondacks: A History of America's First Wilderness* (New York: Macmillan, 1998).

Carl L. Schofield and Chris Keleher, "Comparison of Brook Trout Reproductive Success and Recruitment in an Acidic Adirondack Lake Following Whole Lake Liming and Watershed Liming," *Biogeochemistry* 32, no. 3 (January 1996): 323–37.

Ed Van Put, *The Beaverkill* (New York: Lyons Press, 1996).

Dana Warren, Jason M. Robinson, Daniel C. Josephson, Daniel R. Sheldon, and Clifford E. Kraft, "Elevated Summer Temperatures Delay Spawning and Reduce Redd Construction for Resident Brook Trout (*Salvelinus fontinalis*)," *Global Change Biology* 18, no. 6 (May 2012): 1804–11.

Chapter 11. Connecticut

Mike Beauchene, Mary Becker, Christopher J. Bellucci, Neal Hagstrom, and Yoichiro Kanno, "Summer Thermal Thresholds of Fish Community Transitions in Connecticut Streams," *North American Journal of Fisheries Management* 34, no. 1 (February 2014): 119–31.

Yoichiro Kanno, Benjamin H. Letcher, Ana L. Rosner, Kyle P. O'Neil, and Keith H. Nislow, "Environmental Factors Affecting Brook Trout Occurrence in Headwater Stream Segments," *Transactions of the American Fisheries Society* 144, no. 2 (March 2015): 373–82.

Kirt Mayland and Trout Unlimited, *A Glass Half Full: The Future of Water in New England* (Arlington, Va.: Trout Unlimited, 2006).

Chapter 12. Massachusetts

Brendan Annett, Gabriele Gerlach, Timothy L. King, and Andrew R. Whiteley, "Conservation Genetics of Remnant Coastal Brook Trout Populations at the Southern Limit of their Distribution: Population Structure and Effects of Stocking," *Transactions of the American Fisheries Society* 141, no. 5 (August 2012): 1399–410.

Joseph D. Bergin, "Massachusetts Coastal Trout Management," *Federation of Fly Fishers and Trout Unlimited, Vienna, Virginia* (1985): 137–42.

Daniel C. Dauwalter, Joseph McGurrin, Merry Gallagher, and Steve Hurley, "Status Assessment of Coastal and Anadromous Brook Trout in the United States," *Wild Trout XI* 11 (2014): 192–99.

Doug Fraser, "Sea-Run Brook Trout Vanishing from Cape Waterways," *Cape Cod Times*, November 19, 2015, http://www.capecodtimes.com/article/20151119/NEWS/151119387.

Jerome Van Crowninshield Smith, *Natural History of the Fishes of Massachusetts* (Boston: Allen and Ticknor, 1833).

Erin L. Snook, Benjamin H. Letcher, Todd L. Dubreuil, Joseph Zydlewski, Matthew J. O'Donnell, Andrew R. Whiteley, Stephen T. Hurley, and Andy J. Danylchuk, "Movement Patterns of Brook Trout in a Restored Coastal Stream System in Southern Massachusetts," *Ecology of Freshwater Fish* 25, no. 3 (July 2016): 360–75.

Chapter 13. Vermont and New Hampshire

Biennial Report of the Fish Commissioners of the State of Vermont, vols. 11–18 (Montpelier: Vermont Fish Commissioners, 1892).

Travis W. Blalock and D. R. Lumbert II, *Poverty in Vermont: Reduction and Profiles*: *A Profile of Vermont Residents in Deep Poverty Prepared for the Vermont Child Poverty Council*, PRS Policy Brief 1011-01 (Hanover, N.H.: Nelson Rockefeller Center at Dartmouth College, August 25, 2010).

Ryan Butryn, Donna Parrish, and Donna Rizzo, "Summer Stream Temperature Metrics for Predicting Brook Trout (*Salvelinus fontinalis*) Distribution in Streams," *Hydrobiologia* 703, no.1 (January 2013): 47–57.

Kenneth M. Cox, *Assessment of Trout Cover and Its Relationship to Trout Abundance in the Batten Kill Main Stem and Four Rivers in Reference Watersheds* (Springfield: Vermont Fish and Wildlife Department, 2009).

——— , *Batten Kill Trout Management Plan, 2007–2012* (Montpelier: Vermont Agency of Natural Resources, Fish and Wildlife Department, 2006).

Robert Frost, "Birches," in *Mountain Interval* (New York: Henry Holt, 1916), 37–40.

——, "Mending Wall," in *North of Boston* (New York: Henry Holt, 1914), 11–13.

——, "The Mountain," in *North of Boston* (New York: Henry Holt, 1914), 24–30.

——, "Stopping by Woods on a Snowy Evening," in *New Hampshire* (New York: Henry Holt, 1923), 87.

Ronald J. Hall, Gene E. Likens, Sandy B. Fiance, and George R. Hendrey, "Experimental Acidification of a Stream in the Hubbard Brook Experimental Forest, New Hampshire," *Ecology* 61, no. 4 (July 1980): 976–89.

Peter Matthiessen, *The Snow Leopard* (New York: Viking, 1978).

John Merwin, *The Battenkill: An Intimate Portrait of a Great Trout River: Its History, People, and Fishing Possibilities* (New York: Lyons & Burford, 1993).

Keith Nislow and Winsor H. Lowe, "Influences of Logging History and Stream pH on Brook Trout Abundance in First-order Streams in New Hampshire," *Transactions of the American Fisheries Society* 132, no. 1 (January 2003): 166–71.

Craig Nova, *Brook Trout and the Writing Life: The Intermingling of Fishing and Writing in a Novelist's Life* (Hillsborough, N.C.: Eno, 2011).

Vermont Department of Fish and Wildlife, *The Vermont Management Plan for Brook, Brown and Rainbow Trout* (Waterbury: Vermont Department of Fish and Wildlife, 1993).

R. S. Wentworth, *Biological Effects of Stream Gravel Mining in Vermont* (Waterbury: Vermont Department of Fish and Wildlife, 1987).

W. D. Wetherell, *Vermont River: The Classic Portrait of a Man and His River* (Guilford, Conn.: Lyons Press, 1993).

Chapter 14. Maine

Forrest R. Bonney, *Brook Trout Management Plan* (Augusta: Maine Department of Inland Fisheries and Wildlife, 2001).

——, *Squaretails: Biology and Management of Maine's Brook Trout* (Augusta: Maine Department of Inland Fisheries and Wildlife, 2007).

Eugene J. Conlogue, letter to the editor, *Bangor Daily News*, July 20, 2001.

Robert S. Hogg, Stephen M. Coghlan Jr., Joseph Zydlewski, and Cory Gardner, "Fish Community Response to a Small-Stream Dam Removal in a Maine Coastal River Tributary," *Transactions of the American Fisheries Society* 144, no. 3 (April 2015): 467–79.

David Howatt and David P. Boucher, *Rapid River Fishery Management* (Augusta: Maine Department of Inland Fisheries and Wildlife, 2014).

Frank J. Magilligan, Keith H. Nislow, Boyd E. Kynard, and Alex M. Hackman, "Immediate Changes in Stream Channel Geomorphology, Aquatic Habitat, and Fish Assemblages Following Dam Removal in a Small Upland Catchment," *Geomorphology* 252, no. 1 (January 2016): 158–70.

"Putting Her Money Where Maine's Woods Are," *New York Times*, August 6, 2001, www.nytimes.com/2001/08/06/us/putting-her-money-where-maine-s-woods-are.html.

Catherine Schmitt, "The Salters of Stanley Brook," *Friends of Acadia Journal* 12, no. 1 (summer 2007): 10.

Chapter 15. Labrador

Keith D. Clarke, "An Overview of the Brook Trout Fishery of Newfoundland and Labrador," unpublished presentation to the 144th Annual Meeting of the American Fisheries Society, Québec City, Québec, 2014.

Francesca Fremantle and Chögyam Trungpa, eds., *The Tibetan Book of the Dead: The Great Liberation through Hearing in the Bardo by Guru Rinpoche according to Karma Lingpa* (Boulder, Colo.: Shambhala, 1975).

Johan Hammar, J. Brian Dempson, and Eric Verspoor, "Natural Hybridization between Arctic Char (*Salvelinus alpinus*) and Brook Trout (*S. fontinalis*): Evidence from Northern Labrador," *Canadian Journal of Fisheries and Aquatic Sciences* 48, no. 8 (July 1991): 1437–45.

Norman Maclean, *A River Runs through It and Other Stories* (Chicago: University of Chicago Press, 1976).

Peter Matthiessen, *The Snow Leopard* (New York: Viking, 1978).

Glenn Yannic, Loïc Pellissier, Joaquín Ortego, Nicholas Lecomte, Serge Couturier, Christine Cuyler, Christian Dussault, et al., "Genetic Diversity in Caribou Linked to Past and Future Climate Change," *Nature Climate Change* 4, no. 2 (February 2014): 132–37.

About the Author

Michael K. Steinberg was born in 1965 in St. Louis, Missouri, where he grew up. He completed his bachelor's degree in interdisciplinary studies and master's in geography at the University of Missouri and his doctoral degree in geography at Louisiana State University. He has taught at the University of Southern Maine (1999–2002), University of Hawaii at Hilo (2004–7), and since 2007 at the University of Alabama, where he is a professor of geography, and New College. Steinberg since 2017 has been a coeditor of *FOCUS on Geography*, a publication of the American Geographical Society, and his research has been published in numerous professional and general-interest journals, including *American Fly Fisher:* *The Journal of the American Museum of Fly Fishing, Bonefish and Tarpon Journal, Conservation and Society, Conservation Biology, Economic Botany, Environmental Biology of Fishes, Geographical Review, Mississippi Quarterly, Mountain Research and Development,* and *Professional Geographer.* His other books include *Dangerous Harvest: Drug Plants and the Transformation of Indigenous Landscapes* (Oxford University Press, 2004) and *Stalking the Ghost Bird: The Elusive Ivory-Billed Woodpecker in Louisiana* (Louisiana State University Press, 2008). Beginning in 2023, he is also coeditor of *Southern Culture on the Fly,* an online fly-fishing magazine.